SCIENCE IN ANCIENT INDIA

Science in Ancient India

Edited by
Narendra Kumar

ANMOL PUBLICATIONS PVT. LTD.
NEW DELHI - 110 002 (INDIA)

ANMOL PUBLICATIONS PVT. LTD.
4374/4B, Ansari Road, Daryaganj
New Delhi - 110 002
Ph.: 23261597, 23278000
Visit us at: www.anmolpublications.com

Science in Ancient India

First Published, 2004

ISBN 81-261-2056-8

PRINTED IN INDIA

Published by J.L. Kumar for Anmol Publications Pvt. Ltd., New Delhi - 110 002 and Printed at Mehra Offset Press, Delhi.

Contents

Preface

The earliest indigenous cultures which interest the historican of science are those centred upon Harappa in the Panjab and Mohenjo-daro in Sind, the so-called Indus Valley Civilization. In technology the prominent characteristic is that of standardization: cities built to a uniform plan resembling the layout of a chess-board and of well-fired bricks of a controlled size, and domestic pottery turned from the wheel in specification form and capacity.

Some interesting suggestions might be made from all this, but it would be unwise to resort to idle speculation in the face of such incomplete evidence. Let us be content to add three further practical achievements—the construction of main drains having brick 'manhole-covers', the cultivation of cotton and the manufacture of cotton-cloth, and the working of copperm, bronze, and copper-arsenic alloy. The over-all picture is of a technology standardized.

Ancient India has contributed to human civilization many scientific achievements. In the fields of astronomy, mathematics and medicine they perfected and mastered many systems.

This book portrays the scientific achievements of ancient India in comprehensive way. We have chosen here ten topics to highlight the theme. Topics like— History of Excellence in Science in India; Hindu Calendar; The Mayan and Hindu Calendars: Comparison and Correlation; The Origin and Development of Numerals; The Present Mode of Expressing Numbers; Cosmographical Theories of the Hindu

Astronomers; Aspects of the History of Hindu Astronomy; A Short Chronology of Indian Astronomy; Astronomical Instruments of the Hindus; Greek Influence on Hindu Astronomy: Allegation or Reality?; etc. will put ample light on theme. Scholars and laymen can peep through these to extreme past of India.

The material is gathered from various authoritative sources. As such we owe a deep of sense gratitude to those great scholars. Finally we appreciate the foresight of enterprising Mr. J.L. Kumar, Managing Director, Anmol Publications Pvt. Ltd., New Delhi for publishing this work in superb way.

—Editor

1
History of Excellence in Science in India

Looking at the development of science through the centuries we see an accumulating body of knowledge to which the races of Asia have made their own particular contributions by their own methods of investigation, but we are also confronted by a jigsaw of transmissions which render our interpretation of these discoveries all the more difficult and uncertain. In studying the science of India these transmissions can only be examined at present against a chronology sometimes open to dispute. Yet in spite of this a fascinating story emerges which indeed gains in interest because of its mystery.

I

The earliest indigenous cultures which interest the historian of science are those centred upon Harappa in the Panjab and Mohenjo-daro in Sind, the so-called Indus valley civilization. In technology the prominent characteristic is that of standardization: cities built to a uniform plan resembling the layout of a chess-board and of well-fired bricks of a controlled size,[1] and domestic pottery turned from the wheel in specification form and capacity. These suggest in turn a methodical system in weights and measures. Indeed, a very large number of weights consisting of accurately cut cubes of blended grey chert, which are found to follow the ratios 1 : 2 :

8/3 : 4 : 8 : 16 : 32 : 64 : 160 : 200 : 320 : 640, have been collected at various sites in the Panjab, Sind, and South Baluchistan, and probably the Makran.

This system of weights is unique in the ancient world. It is unfortunate that no numerals appear on any of these weights—as in the case of the Harappa script, which is pictographic and has yet to be deciphered, the mathematical achievements of the people remain largely a mystery. Clearly there was a considerable merchant class, through whom a commercial arithmetic developed, but academic achievements, like those of contemporary Egypt, were no doubt both channelled and circumscribed. However, it will be seen that the above system of ratios may be based upon 16, an important number[2] in ancient Indian numerology, and that certain others may be successively obtained by doubling or halving; also of interest is the use of fractional thirds and the development of a decimal form in the higher numbers. Further study of the metrology of the Indus valley civilization does, in fact, reveal decimal divisions of length, e.g. the use of a 'foot' of 13.2 inches divided into tenths. It is clear from the planning and architecture of the cities that there was a competent knowledge of simple geometry and surveying based upon two units of length, a 'foot' of about 13.2 inches and a 'cubit' of about 20.6 inches.

Some interesting suggestions might be made from all this, but it would be unwise to resort to idle speculation in the face of such incomplete evidence. Let us be content to add three further practical achievements—the construction of main drains having brick 'manhole-covers', the cultivation of cotton and the manufacture of cotton-cloth, and the working of copper, bronze, and copper-arsenic alloy. The over-all picture is of a technology standardized through several centuries by an inflexible and authoritarian regime.

II

The Aryan invasions of northern India (*c.* 1500 B.C.) may

be said to mark the end of the Indus valley civilization. From henceforth the growth of Indian science is to be influenced by the speculative and philosophical mind, to become richer in generalization, to transcend the limited technology of Harappa, Mohenjo-daro, and Chanhu-daro. In the hymns of the *Rig Veda* is to be found the first account of the way of life of the Indo-European conquerors, their recognition of and devotion to one supreme cause, their realization that behind the phenomena of the natural world, which appear shifting and changeable, there is a constant principle (*rita*) or order in events. Piggott[3] has shown that the war-chariot (*ratha*) of the *Rig Veda* had a central pole and yoke harness, the so-called throat-and-girth harness, not only unpleasant for the horse but most inefficient mechanically, yet nevertheless common to the regions of Indo-European colonization, for example Homeric Greece and Celtic Britain. Despite the gradual development of philosophy the personification of the primal forces of nature in, for instance, the god of the sun, Surya, or the god of fire, Agni, continued. Sacrificial altars, at first mere heaps of turf, evolved into elaborate designs demanding arithmetical and geometrical calculations.

Vedic literature, broadly considered, tells us only fragmentary information concerting the early stages of Hindu science. The wisest procedure is to examine the whole evaluation in the light of certain terminal writings such as the *Kautillya Arthasastra*, the *Sulva Sutras*, the *Caraka Samhita*, and ultimately the *Surya Siddhanta*, which present us with an established body of knowledge or doctrine. Assuming tentatively a date of 1500-1400 B.C. for the *Rig Veda* we are confronted by a vast ocean of time from which small 'islands' appear at irregular intervals to provide a few bearings, and firm ground is denied us until after the dawn of the Christian era in Europe.

Perhaps the earliest source dealing exclusively with astronomy is the *Jyotisha Vedanga* from this work, the text of which is corrupt and condensed in form, and may date from

500 B.C., one learns the rules for calculating the position of the new and full moon amongst the 27 *nakshatras*, and of the *ayanas* which fall in cycles of 5 years each of 366 days. In 5 solar years were 67 lunar months, so that if these are taken as equivalent to 62 synodic months, then a year of 12 months may be retained if the 31st and 62nd months are omitted from each cycle. This ancient system of lunar-solar reckoning was widely used in India and occurs also in Jaina literature. An earlier statement on the *nakshatras* in the *Taittiriya Samhita*[4] (of the same period as the *Brahmanas* or priests' books of ritual) gives a complete list of their names, which must have been well established by the eighth or seventh centuries B.C.,[5] and certainly existed in the sixth century at the very latest. Of the three independent systems of astronomical reference used in antiquity, the decans of Egypt, the zodiac of Mesopotamia, and the lunar mansions, i.e. the 27 or 28 positions occupied by the moon in one sidereal rotation, the last seems to have appeared in India as *nakshatras* and in China as *hsiu* at about the same time.[6] There is no evidence to show that at this early period the development could have been anything but independent: but if one is looking for a common origin there is just the possibility that moon-stations could have arisen from the old Babylonian astronomy of *c.* 1000 B.C. and the conception been diffused through Iran: their reappearance as *al-manazil* among the Arabs, the heirs of Sassanian Iran, in pre-Qur'anic times is suggestive.

Early Indian cosmology is generally based upon the square and cube—a quotation from C. P. S. Menon[7] summarizes the basic concept adequately: There is first of all the earth based on a square, with a corner towards the south, and shaped like a pyramid, with a number of successive homocentric square terraces rising up to a point (or rather, to a small square): on the top of this is the mount Meru, a pyramid widening out as it rises, at a small angle to the vertical; round this lie the orbits of the sun forming homologous squares on a horizontal plane; above the sun's plane is that of the moon with similar orbits.

We may imagine above this were the planes of the different planets at increasing heights, as described in the Vishnu-Purana (of the Hindus); if these were also originally square orbits, we should have the original conception of the orbits of the planets as forming the successive terraces of a pyramid representing the heavens.

Associated with this early Meru cosmology were a series of numbers, such as 4, 12, 28, 60, obtained through sub-division of the square, or rectangle. The system was dictated by mathematical rather than by astronomical requirements. Thus a 'square' orbit of 28 could be represented by placing unit squares around the periphery of an original square of side 6 (containing 36 unit squares) and adding in the unit square at each corner, i.e. (4×6) + 4, giving a geometrical picture of the alighting stations of the moon.

Later Meru cosmology, which is to be found in the Jaina texts adopts the circle as the basic form. This radical departure occurs in the *Surya Prajnapti* (perhaps 200 B.C.) where the earth is represented as a circular disc with Mount Meru as its centre and the pole-star directly above. Surrounding the earth are seven concentric oceans and continents, whilst coplanar rotations of the planets are from east to west around Mount Meru. In the *Jambudvipaprajnapti* a further step is taken when the detailed geometry and associated calculations of this circle are made on the basis of the ratio circumference: diameter= $\sqrt{10}$. Numerical results are obtained for lengths of arcs, sagittae, and segments of chords, and in certain cases quadratic solutions are required. Though the history of Jaina canonical literature is such that it is difficult to date, some useful papers have already appeared on the mathematical and astronomical aspects.[8]

Maha Meru, or Sineru, also occupies the central position in the Buddhist universe as described in the Pali Books[9] *Sara Sangaha, Visuddhi Magga, Satta Suryuggamana Suttanta,* and *Jinalankara*. Only one half of it is visible above the level of the ocean, whilst between Meru and the outermost circle of the system, the ridge of rock called Sakvala, there are seven circles of rock, and seven oceans the 'waters' of which are so attenuated that the feather of a peafowl could not float in them. This picture has been much elaborated, so that to proceed further would only involve us in vast numbers and fanciful speculations having no relation to fact, curious and interesting though these may be. The most important achievement of Buddhist thought from our present standpoint is the law of causation, which was initially concerned mainly with human conduct but has since interested philosophers of science.[10]

Whilst dealing with Jaina and Buddhist cosmology we must mention the doctrine of *paramanus,* better known as the atomic theory. Atoms could be grouped together to form molecules, and whilst the atoms envisaged by the Jains and Vaiseshikas were eternal, those of the Buddhists, being included in a phenomenalism view of nature, appear and disappear by cycles. In the *Kevaddha Sutta* appended to the *Digha Nikaya*[11] we also find that all matter is ultimately formed from the four elements, fire, air, water, and earth. To these is added, in the *Maitri Upanishad,* a fifth 'element', non-material and all-pervading. The fully developed atomic theory in the *Abhidharma Hridaya,* which was translated into Chinese in the third century A.D., represents the refinement of a doctrine which had existed for several centuries, but McGovern is inclined to the view that both the Jaina and Buddhist versions are not original, but probably derived from the Vaiseshikas. Whatever the source, it is Indian and seems independent of the Greek theories of Leucippus and Democritus.[12]

To learn more of the worldly activities of the Indian peoples in the last few centuries B.C. one must seek information other than that supplied by the many religious and metaphysical writings, and it is fortunate that a text of the *Kautiliya Arthasastra,*[13] a unique work on statecraft attributed to Kautilya, prime minister of the Mauryan emperor

Chandragupta who ruled from about 321 to 300 B.C., was discovered in 1909. Probably elaborated from the Mauryan original and containing references to the economics and technology of a mean date *c.* 100 B.C., it is a store of information on land and sea communications, agriculture and irrigation, ores and mining, plants and medicine, and especially mechanical contrivances or *yantras*.[14] Engines of war and *yantras* of architecture are mentioned, with emphasis mainly on the former; a whole chapter is devoted to armoury, and the *yantras* are classified as stationary or mobile; and even though their detailed operation can often only be inferred it is clear that the main descriptions refer to siege warfare, the role of elephants in war, catapult devices for hurling projectiles, and incendiary missiles, the composition of the inflammable materials used indicating the absence of gunpowder. In the furtherance of cultivation, irrigation canals and large artificial reservoirs such as the lake of Girnar in Saurashtra (Kathiawar) were constructed, and these had their counterpart in the massive 'tanks' of Ceylon, first seen in the Abayawewa of King Panduwasa built near the capital Anuradhapura as early as 504 B.C.[15] A successor, King Pandukabhaya, had constructed in the latter half of the fifth century B.C. two further 'tanks', the Jayawewa and the Gamini, in the same region, and such stupendous activity in irrigation works went on, as in India, into the medieval period. One cannot but marvel at these splendid monuments, some of which were 40 miles in perimeter, and the phenomenal agricultural progress achieved during half a millennium from the accession of Vijaya, the first of the Great Dynasty, nor fail to record the scholar's indebtedness to George Tumour, who sought out the 'tika' with the help of the Buddhist priest Gallé and through inconceivable difficulties produced an English version of a third of the *Mahavamsa* in 1836, which rendered our historical knowledge of these early public undertakings possible.

The machines of the period of the *Arthasastra* and of the early centuries A.D. were either useful devices of everyday life such as the *variyantra*, probably a revolving water-spray for cooling the air, mentioned by the poet Kalidasa in his *Malavikagnimitra;* or automata and toys such as those described by Hero of Alexandria, Philo of Byzantium, and Vitruvius.[16] References to *Yavanas*, who were often engineers, in early Tamil literature, and commercial intercourse between southern India and the West, especially in Augustan times, suggest acquaintance with Greek and Roman ideas. Though Bhoja in his treatise *Samaranganasutradhara* gives many technical properties of machines neatly classified, refers to the use of toothed wheels in the operation of types of merry-go-round, and has a general statement concerning the magnitudes of effort and load in a machine, we do not find scientific laws as in Greek mechanics. Perhaps most remarkable is the absence from the Jaina physics of a concept of force, action and change arising out of time, but, as in the case of the atomic theory, this must be considered in terms of the contemporary philosophy; there is error in reading too much into early texts.

Ancient Hindu mathematics shows an early interest in large numbers expressed in powers of ten, in the nature of numbers and their factors, and in the division of time into its smallest units. These large powers occur in the Vedic *Samhitas, Brahmanas*, and *Sutras*, in the epics *Mahabharata* and *Ramayana*, and in the *Lalitavistara* (where 10^{53} is given). Of particular interest is the *Satapatha Brahmana*, which lists all the factors of 720 as far as 24, and after stating that 360 nights and days contain 10,800 *muhurttas* proceeds by four successive multiplications by 15 to reach the ultimate *pranas* or breathings. The occurrence of the word *rasi* (a heap) in the *Chandogya Upanishad* recalls the use of the same concept by the Ancient Egyptians, and is clearly the humble origin of what was later to become the burden of many a schoolboy, standing for the unknown quantity *x*.

Of the greatest importance to the historian of mathematics are the *Sulva Sutras*,[17] which form part of the *Kalpa Sutras* and deal with the construction of sacrificial altars used in Vedic ritual. As terminal writings they summarize the knowledge of several preceding centuries and provide an excellent picture of the achievements of Hindu geometry prior to the mathematics of the Jaina sect; furthermore, when temple worship replaced the old rites of the *agni-cayana,* this geometrical tradition lapsed[18] and was subsequently superseded by the growth of analysis for which Hindu mathematicians are justly renowned. From the mass of literature which must have been the prerogative of the priesthood seven *Sulva Sutras* have survived and of these three are especially valuable—those of Baudhayana, Apastamba, and Katyayana.[19] They deal with such matters as the construction of squares and rectangles, the relations of the sides to the diagonals, the construction of equivalent squares and rectangles, the construction of equivalent squares and circles, the construction of triangles equivalent to squares and rectangles, and the construction of squares equal to two or more given squares or equal to the difference between two given squares. In this connection we may note two interesting formulae, those giving the diagonal of a square and the squaring of the circle. Thus, according to Baudhayana and Apastamba, to obtain the *dvi-karani* or diagonal[20] 'Increase the measure by its third part, and again by the fourth part (of this third part) less the thirty-fourth part of itself (i.e. of the fourth part).' This gives a value for $\sqrt{2}$ of

$$1+\frac{1}{3}+\frac{1}{3.4}-\frac{1}{3.4.34} \text{ or } 1.4142156...$$

diverging from modern calculation only in the sixth place of decimals.

Baudhayana says: 'If you wish to square a circle, divide its diameter into eight parts; then divide one of these parts into twenty-nine parts and leave out twenty-eight of them; and also the sixth part (of the previous division) less the eighth part of this (last).'[21] A relation between the radius of the circle (r) and the side ($2a$) of the equivalent square is finally obtained in the form:

$$a = r - \frac{r}{8} + \frac{r}{8.29} - \frac{r}{8.29.6} + \frac{r}{8.29.6.8}$$

In the construction of altars (*vedi*) requiring numbers of bricks of differing sizes in various layers, e.g. in the falcon-shaped fire altar, we see the origin of those indeterminate problems which form a notable part of later Hindu algebra. Bibhutibhusan Datta, after examining Jaina canonical literature and the commentaries of Kapardisvami and Karavindasvami, inclines also to the view that the irrationality of $\sqrt{2}$ was understood in the time of the Sulvas.

Archaic Hindu medicine in its earliest context is to be found mostly in the hymns of the *Atharva Veda*, and the Vedic term *bheshaja*, used to denote medicinal charms, which occurs also in the *Avesta* as *baesaza* or *baesazya*, suggests a common Aryan origin. It is a 'psychosomatic' approach to healing, part of a philosophical system, a scheme in which the lay physician and the priest perform their respective roles in controlling the ills of the body and of the soul. Native pre-Aryan lore and practice were absorbed. Mohenjo-daro, which had the finest public health facilities in the ancient East, could boast bathrooms and a drainage system, and no doubt influenced personal hygiene. The deity Dhanvantari, custodian of the elixir of immortality, became the fount of wisdom for virility and duration in life (*ayurveda*) and the remedies (*bhaishajya*) to ensure these. He figures in Susruta as the divine authority in medicine. We cannot, therefore, compare such a system with contemporary medical education, drawing as it does upon several of the exact sciences, but we can perceive the beginnings of certain features in general practice and in surgery which are still with us; thus, we find the practitioner teaching his pupil by personal example, a blend of experience and tradition with fresh observation and speculation, whilst specialized surgical equipment in common use by the first century A.D. consisted of twenty types of knives and needles '(*sastra*), thirty probes (*salaka*), twenty tubular instruments, and twenty-six articles of dressing (*upayantra*).[22]

Its theoretical basis being metaphysical, Hindu medicine was restricted by tradition and by isolation from other sciences. Life originated in the primal waters, man the microcosm was inevitably moulded by the forces of the macrocosm;[23] the types of physiognomy are the result of specific incarnations. In the *Puranas* fever is a demon, offspring of indigestion—the commonest cause of illness. Since illness in general could be attributed either to a disturbance in the organism or to the entry of super-human forces, then, should the physician fail, prayers must be recited and offerings made to propitiate the intruder. The human body was maintained in a state of health by the three humours, phlegm, gall, and wind (or breath) in their correct proportions.[24] These proportions could be achieved by proper diet, an important consideration in a trying climate. The humours were forms of the life-energy and corresponded to divine forces or agents in the macrocosm, i.e. outside the body; thus phlegm, cool and heavy, which resided in the chest and lungs, was associated with the moon. Hindu interpretation gave wind prime significance among the humours, since it appeared to govern the dynamics of the body; from the pre-Aryan Yoga to the Vedanta philosophy there developed a theory of winds (*vayu*) or manifestations of the life-breath.[25] *This is a theory of breathing which ignores the lungs.* In fact, the lungs (*kloman*) and the palate were *regarded* by Caraka[26] as the source of the vessels which carry water through the body. Existing alongside Caraka's description of the vascular system was that of Susruta, postulating the navel as the source; of the tubular vessels (*sira*) radiating therefrom, each of the humours (now four in number) occupied 175. So in all: 'There are seven hundred tubular vessels. As a garden or a rice field is irrigated by a system of canals carrying water, the body, by means of these vessels, is moistened and maintained;... the root from which they spring is the navel... the navel is surrounded by the tubular vessels, as the hub of a wheel is surrounded by spokes.'[27]

Disregard of proper diet leads to a disturbance of the balance of the humours; these become incensed, and overflowing their normal channels invade the domain of others, thereby causing disease. The basis of dietetics and pharmacology was the Hindu theory of the six essences (*rasa*), which appear to correspond to the Greek *glyky, liparon, stryphnon, halmyron, pikron,* and *drimy.*[28] Tradition had established an elaborate doctrine of correspondence between the essences (qualities or flavours) in certain foods and the specific substances in the humours. A vast pharmacopoeia enshrined traditional remedies. In the treatment of the patient an all-powerful arcanum was blended from various herbs, each contributing specific healing properties, and these material properties were reinforced by supernatural powers invoked by the brahman practitioner who claimed a special relationship with the twin horsemen and divine physicians, the Asvins. Honey possessed unusual healing virtues, being associated with *amrita,* the elixir of immortality.

Another significant feature of Hindu medicine was *the absence of any attempt to recognize diseases of the brain.* One would not have expected much progress in this study in ancient times in any case, but the reason for such neglect is to be found in the assumption that the centre of consciousness, thought, and feeling is the heart, a generalization which is implicit also in the writings of Homer.[29] Caraka[30] certainly mentions insanity, but it is covered by the general explanation of the overflowing of 'incensed elements', in this instance into the special vessels carrying the 'mind-stuff', or by the entry of demons.

Among the diseases mentioned in Vedic medical texts are diarrhoea (*asrava*), fever (*takman*), dropsy (*jalodara*), consumption (*balasa, yakshma*), tumour (*akshata*), abscess (*vidradha*), leprosy and certain skin diseases (*kilasa*), and congenital diseases (*kshetriya*). Dropsy was sent by Varuna, the god of the primal waters. Jaundice, for which Caraka later records diagnosis and treatment,[31] is characterized by the presence of the demon who causes yellowness (*hariman*).

On pursuing inquiry to the borders of the realm of legend the historian gleans one indisputable fact, namely, that the fount of ancient Hindu medical wisdom was the oral teaching of Punarvasu Atreya. According to the Buddhist *Jatakas* a physician Atreya taught at Takshasila (Taxila) in the age of Buddha. It appears that six pupils of Atreya first set down this wisdom, in encyclopedic form, but of these versions only two, those of Bhela and Agnivesa, have survived. A defective manuscript of *Bhela Samhita*,[32] discovered in south India, reveals the same tradition as is to be found more fully expounded in the *Caraka Samhita*, which is the final form of the compilation of Agnivesa and is our best source of Hindu medical knowledge as it existed in the last few centuries B.C. Caraka is generally believed to have been the court physician to King Kanishka at Peshawar in the first or second century A.D.

H. R. Zimmer, on the evidence of the Bower manuscript,[33] suggests a possible third, school parallel with Bhela and Agnivesa. This oldest extant medical manuscript, discovered by Lieut. A. Bower in 1890 in a Buddhist monument at Kucha, Chinese Turkestan, has been dated on the basis of palaeography in the second half of the fourth century A.D. but contains material deriving from several centuries earlier, and not only has certain chapters corresponding to those of Caraka but almost identical pharmaceutical formulae. The work is interesting in that no reference is made to Caraka; it is also significant in showing the penetration of Hindu medicine into Central Asia.

Ayurveda (the science of longevity) as set forth in Caraka makes no mention of surgery, being solely the province of the physician. The development of surgery is initially attributed to the genius Susruta who may have taught and practised in Kasi (Varanasi, Banaras). He incorporated surgery into the general field of medicine, advised a wide training and experience gained under several teachers, stressed the importance of surgery in the study of anatomy (which was the major weakness in Hindu medical knowledge), and attempted a stricter classification of existing data[34] which still resided in separate monographs marred by confusion and repetition. With Susruta also there ended the specialized tradition of elephant medicine.[35]

Hindu tradition made an eight-part division of the field of study, broadly in respect of (a) diseases, their diagnosis and treatment, and (b) the means of healing in relation to the whole man, the philosophical and ethical approach. Thus under (a) we find illnesses requiring surgery (*salya*) and the science of obstetrics; diseases of the eye, ear, nose, and throat (*salakya*); diseases due to the disturbance of the humours which involve the therapy of the whole organism; mental and other disturbances of demoniacal origin; pediatrics, i.e. children's diseases, caused by demons; and finally, three aspects of *ayurveda* —medicinal drugs *(agada)* and antidotes, elixirs of life (*rasayana*), and virility (*vajikarana*). More widely (b), we consider the organism (*sarira*), its moral and physical health (*vritti*), the origins of disease, and the nature of pain and illness in terms of the balance of the humours, treatment or action (*karman*), the consequences of treatment, the influence of time (*kala*) in respect of the age of the patient or perhaps the seasons, and lastly, the professional conduct of the agent or physician, his diagnosis, his methods and instruments (*karana*). Emphasis was laid upon the preventative aspect and early treatment.

There is no reference to hospitals in the ancient Hindu medical literature, but they evolve with the spread of Buddhism. The second Rock Edict (*c.* 256 B.C.) of the Mauryan

Emperor Asoka celebrates the beginnings of social medicine, whilst Ceylon, by the fourth century A.D., could boast some hospitals, and a medical service; by royal command each physician served the villages, and veterinary officers tended the king's elephants and horses. Ample evidence of the treatment of out-patients in dispensaries occurs in the *Sangam* literature of southern India.[36]

III

We conveniently leave the ancient world via the *Siddhantas,* the astronomical treatises which in themselves exhibit a transition from the *Paitamaha Siddhanta,* which retains the Vedanga astronomy, to the *Surya Siddhanta* of A.D. 400,[37] which largely establishes the form of native astronomy for the duration of the Middle Ages.[38] Varahamihira, *c.* A.D. 505 summarized in his *Panchasiddantika* the five *Siddhantas* entitled *Paitamaha, Vasishtha, Paulisa Romaka,* and *Surya,* though his version of the last-named indicates that gradual changes in the text of this, the most important Siddhanta, must have occurred subsequently. K. S. Shukla lists as a minimum twenty-eight commentaries on it by known authors, mostly in Sanskrit but two in Telugu, reaching to the early eighteenth century, together with at least seventeen works based essentially upon its theory; his recent edition[39] includes the commentary of Paramesvara (A.D. 1432) written in Kerala in south India.

Much discussion has centred around the transmission of Greek astronomical ideas to India during the first four centuries of the Christian era.[40] This period coincides with that of the growth of the Siddhanta literature, and the *Romaka Siddhanta* especially shows signs of Greek influence, an influence which is notably present in the terminology of astrological writings such as the *Brihajjataka and Laghujataka* of Varahamihira. It is also the period of close commercial intercourse between imperial Rome and the coasts of Kerala and Tamilnadu, embracing both the Augustan age and the Sangam age.[41] Tamil poems of the latter make frequent references to *yavanas*,[42] who were Westerners familiar with Hellenistic science and Roman technology, and who assisted in the design and construction of *yantras*, especially engines of war. More or less contemporary influences of Greek and Roman craftsmanship are evident in coinage and in sculpture in northern India.[43] Nor should one overlook the intellectual contacts of the Gupta Empire with Sasanian Persia, where there was some study of astronomy.[44]

The Tamil tradition in respect of astronomy is especially significant. By the study of two Greek papyri of the Roman period, which seem to have been written c. A.D. 100 and *c.* 250 respectively,[45] and of the Tamil methods of reckoning time,[46] O. Neugebauer has revealed the persistence of the Babylonian methods of the Hellenistic period up till recent times in southern India. Starting from the work of Lt. Col. John Warren of 1825, entitled *Kala Sankalita: a Collection of Memoirs on the various modes according to which the natives of the southern parts of India divide time,* he has found exact numerical parallels between the planetary theory of Varahamihira's *Panchasiddhantika* and that of Seleucid cuneiform tablets of *c.* 200 B.C., the intermediary phase being the Roman transmission of the earlier form of Greek astronomy, which owed much to Babylonia, into southern India, probably mainly through the medium of the astrological literature. Calculations were made directly for the prediction of astronomical phenomena through certain numerical schemes, and we may call this tradition 'arithmetical'.

With the *Almagest* of Ptolemy we reach a new and more highly developed process of mathematical deduction from geometrical models, which represent in themselves the complicated motions of the planets in terms of epicycles or eccentrics. The *Surya Siddhanta,* in addition to the use of terminology and units of Greek origin, employs epicyclic models in its planetary theory. But it is significant that it does not include Ptolemy's refinement of the lunar theory or the 'punctum aequans' of Ptolemaic planetary theory.[47] Neugebauer has now succeeded in reducing the essential features of the planetary theory of the fully developed *Surya Siddhanta* to modern terms; thus he says: 'The leading idea is the following. The planet moves on an epicycle of radius r which is carried on a circle of radius R and eccentricity e, the 'deferent' around the observer. Thus we are dealing with two variables, the 'mean distance' α of the centre of the epicycle from the apogee of the deferent, and the 'anomaly' γ which determines the position of the planet on the epicycle. The problem now arises to tabulate this rather complicated function of α and γ.'[48] Into this Greek geometrical system the Hindus injected the important concept of the sine of an angle, thus initiating a second tradition which we may call 'trigonometrical'. Both the *Paulisa* and *Surya Siddhantas* contain a table of sines (*jya*).

By remembering that there is a Dravidian component as well as an Aryan one, that there are, in fact, two traditions, the arithmetical and trigonometrical, a proper perspective on the development of Indian astronomy may be obtained in terms of the data furnished by both Tamil and Sanskrit sources.

The chapters of the *Surya Siddhanta*[49] deal with (I) the mean motions of the planets, (II) the true positions of the planets, (III) direction, place; and time, (IV-VI) the nature of eclipses, (VII) planetary conjunctions, (VIII) asterisms, (IX) heliacal risings and settings, (X) the rising and setting of the moon, (XI) 'certain malignant aspects of the sun and moon' treated in part astrologically, (XII) cosmogony, geography, and the 'dimensions of the Creation', (XIII) measuring instruments, such as the armillary sphere, clepsydra, and gnomon, and (XIV) different ways of reckoning time. The geocentric system of the universe is assumed, and longitudes reckoned from the prime meridian through the ancient city of Ujjayini. In spite of the imported Greek ideas, the *Surya Siddhanta* is characteristically Indian in its reaction to this alien knowledge;

it absorbs what seems to be fitting and treats it in accordance with its own way of thinking, retaining at the same time certain elements of tradition, e.g. the established feature in time-reckoning, the *tithis,* which are constant and equal to thirtieths of a mean synodic month,[50] a division which has also been traced in Babylonian lunar and planetary texts; and the explanation of the irregularities of planetary motion in terms of the winds, which were a significant feature in ancient Hindu philosophy concerning the natural world. Thus, in the *Satapatha Brahmana* the smallest units of time were *the pranas* or breathings, whilst in the *Surya Siddhanta*[51] we find that 'Forms of Time', of invisible shape, stationed in the zodiac (*bhagana*), called the conjunction (*sighroccha*), apsis (*mandoccha*), and node (*pata*), are causes of the motion of the planets.

The planets, attached to these beings by cords of air, are drawn away by them, with the right hand and left hand, forward or backward, according to nearness, toward their own place.

A wind moreover, called provector (*pravaha*) impels them toward their own apices (*uccha*); being drawn away forward and backward they proceed by a varying motion.

The so-called apex (*uccha*), when in the half-orbit in front of the planet, draws the planet forward; in like manner, when in the half-orbit behind the planet, it draws it backward.

When the planets, drawn away by their apices (*uccha*), move forward in their orbits, the amount of the motion so caused is called their excess (*dhana*); when they move backward, it is called their deficiency (*rina*).

Highly regarded and widely disseminated, the *Surya Siddhanta* had a profound influence on the course of medieval Hindu astronomy. According to Sumati (*c.* A.D. 800), whose work was known both in Nepal and in Kerala, and who wrote his *Sumati Tantra* and *Sumati Karana* on the basis of the earlier version of the *Surya Siddhanta*, it provided the essential elements used by Nepalese astronomers in their construction of the Hindu calendar. Evolving during the period between A.D. 628 and 966, the later version gained greatly in popularity, especially in the twelfth century, when Bhaskara II quoted from it and Mallikarjuna Suri wrote commentaries on it, first in Telugu then in Sanskrit.

The more important Hindu astronomers, with the approximate dates at which they flourished, were Aryabhata I (A.D. 499), his pupil Latadeva (A.D. 505), Varahamihira (A.D. 550), Brahmagupta (A.D. 628), Bhaskara I[52] (a contemporary of Brahmagupta and a disciple of Aryabhata I), Lalla[53] (A.D. 748), Manjula (A.D. 932), Aryabhata II[54] (A.D. 950), Sripati[55] (A.D. 1039; author of the *Siddhanta Sekhara* and *Ganita Tilaka*), and Bhaskara II (A.D. 1150), who wrote the important treatise *Siddhanta Siromani.* Since most of them also made significant contributions to mathematics we shall meet them again in the ensuing pages.

Hindu mathematics is undoubtedly the finest intellectual achievement of the subcontinent in medieval times.[56] It brought alongside the Greek geometrical legacy a powerful method in the form of analysis, not a *deductive* process building upon accepted axioms, postulates, and common notions, but an intuitive insight into the behaviour of numbers, and their arrangement into patterns and series, from which may be perceived *inductive* generalizations in a word, algebra rather than geometry. This native power has, fortunately, survived to modern times; Srinivasa Ramanujan (1887-1920), in whom resided a phenomenal memory allied with great facility in calculation, knew the positive integers 'as personal friends'. The quest for wider generalization beyond the limitations of pure geometry led the Hindus to abandon Ptolemy's method of reckoning in terms of chords of a circle and to substitute reckoning in sines, thereby initiating the study of trigonometry. It is to the philosophical mind of the brahman mathematician engrossed in the mystique of number that we owe the origin of analytical methods. In this process of abstraction two particularly interesting features emerged, at the lower level of achievement the perfection of the decimal system, and at the higher the solution of certain indeterminate equations.

The first Hindu algebraist was, Aryabhata I, of whose mathematical work we have only thirty-three *slokas* which form a section of the astronomical writing *Aryabhatiya*.[57] He had a prolonged influence in both astronomy and mathematics, a commentary, *Bhatadipika*, upon the *Aryabhatiya* being written by Paramesvara as late as 1430. The condensed form in which the mathematical knowledge of Aryabhata appears serves mainly as a criterion of the state of the subject at the end of the fifth century A.D., but certain topics clearly emerge, e.g. square and cube roots, simple areas and volumes, the simpler properties of circles, sines, gnomon problems, arithmetical progressions, factors, and simple algebraic identities; whilst π is stated as $3\frac{177}{1250}$ (i.e. 3.1416). Algebra is now defined as a separate study (*bija*) and there is given a general solution in whole numbers of the indeterminate equation of the first degree. Sections 3 and 4 of the *Aryabhatiya* deal respectively with time reckoning (*kalakriyapada*) and spherical astronomy (*golapada*), the latter treatment ultimately bearing its full fruition in the spherical trigonometry of the Muslim astronomers Abu Raihan al-Biruni[58] and Nasir al-Din al-Tusi.[59]

Whilst Aryabhata I excelled as an observer and in the classification of astronomical data, Brahmagupta was stronger as a mathematician. Brahmagupta is noted for his *Brahmasphuta Siddhanta* (A.D. 628),[60] where in the twelfth and eighteenth chapters may be found important mathematical developments, and his *Khandakhadyaka*[61] (665), with its supplement *Uttara Khandakhadyaka*. The work on cyclic quadrilaterals (i.e. four-sided figures with their angles on the circumference of a circle) is an interesting example of the achievement of the Hindus in the geometry of lines or boundaries, in terms of which they tended to think rather than in terms of angles. Brahmagupta showed that:

1. If the sides of a cyclic quadrilateral are of lengths a, b, c, d, and its semi-perimeter is s, then its area[62] is

   $$A = \sqrt{(s-a)(s-b)(s-c)(s-d)}$$

2. If the diagonals of the same quadrilateral are of lengths x and y, then the relations between these diagonals and the sides of the quadrilateral are expressed by 'Brahmagupta's Theorem'.

   $x^2 = (ad + bc)\,(ac+bd) \div (ab + cd)$

   $y^2 = (ab + cd)\,(ac+bd) \div (ad+bc)$

3. If a, b, c and p, q, r are the sides of two separate right-angled triangles, such that $a^2 + b^2 = c^2$, $p^2+q^2 = r^2$, then if we make a quadrilateral of which the sides are the products ar, cq, br, and cp, this quadrilateral, called 'Brahmagupta's Trapezium', will be cyclic and its diagonals will intersect at right angles.[63]

Using the newer concept of sines of angles, not angles, Brahmagupta, in the first stanza of his chapter in the *Khandakhadyaka* on the rising and setting of planets, gives the formula which is now expressed as

$$\frac{a}{\text{Sin}\angle \text{A}} = \frac{b}{\text{Sin}\angle \text{B}} = \frac{c}{\text{Sin}\angle \text{C}}$$

where a, b, c are the sides of the triangle ABC.

The astronomical writings of Brahmagupta were known in western India at the time of the Muslim invasion of Sind (A.D. 712) and also to Abu Raihan al-Biruni on his Indian journey some three centuries later, and there is little doubt that they were one of the media through which Hindu astronomy and mathematics passed to the Arabs during the 'Abbasid caliphate.[64] Of a practical nature, the Arabs transmitted to the West the so-called Hindu numerals and decimal system and the simpler algebraic and trigonometrical processes, but ignored the use of negative quantities and the higher algebra of indeterminate equations which they do not appear to have understood.

Medieval Hindu mathematics may be conveniently divided into two provinces of study, *patiganita*[65] and *bijaganita.* The former comprised mainly arithmetic and mensuration; geometry, which had earlier been a distinct discipline in the *Sulva Sutras,* was now widened in scope and assimilated into mensuration. Perhaps the earliest example of this kind of presentation is to be found in the Bakhshali manuscript,[66] which is written in old *Sarada* characters on seventy folios of birch-bark, and was unearthed from a mound in the Peshawar district of north-west India in May 1881. Uncertainty surrounds its age; the *mathematics* has been dated as early as the third century A.D. and the manuscript itself as late as the twelfth. *Bijaganita,* which was concerned mainly with the solution of algebraic equations, embraced problems in which there were more unknown quantities than there were equations with which to find them. This study of indeterminate equations was a notable feature of both Hindu and Chinese mathematics during the medieval period.

In India indeterminate analysis reached its zenith in Bhaskara II. He described the solution of the first-degree equation in terms of the pulverizer (*kuttaka*),[67] a quantity such that, when it is multiplied by a given number, and the product added algebraically to a given quantity, the sum or difference is divisible without remainder by a given diviser; that is, he obtained whole-number values of x and y which satisfy the equation $ax \pm by = c$. Indeterminate equations of the second degree in the forms

$$ax + by + c = xy$$

and

$$ax^2 + c = y^2$$

had already been investigated by Brahmagupta, but the solution of the general equation

$$ax^2+bx+c = y^2$$

by the *chakravala* or cyclic method, was effected by Bhaskara II in a manner which has perpetuated his name for all time in the history of the theory of numbers.[68] It is salutary to remember that Bhaskara II made these advances around the middle of the twelfth century; independent European investigations of the seventeenth and eighteenth centuries did not reach completion until about 1770, with the work of Euler and Lagrange.

The delightful *Lilavati* and *Bijaganita,* which form part of the *Siddhanta Siromani* of Bhaskara II, have been widely used since their composition. A Persian version *of Lilavati*[69] appeared in 1587 on the orders of Akbar and one of *Bija ganita* in 1635 for Shah Jahan. *Bijaganita* is interesting in that it contains a demonstration of the theorem of Pythagoras which is distinct from that of Euclid 1, 47. The deductive proof in Euclid represents the squares drawn externally upon the three sides of a right-angled triangle, whereas the figure in Bhaskara, which may well derive ultimately from the Chinese *Chou Pei Suan Ching* of the Han period, is not part of a deductive system but is simply a practical demonstration which shows how two given squares may be so cut that the parts fit together again to form a third square.[70] *Lilavati* was later rivalled by the *Ganita Kaumudi,* composed in 1356 by Narayana, a work notable for its treatment of magic squares.[71] Indian interest in magic squares is reflected in Siamese mathematics of the seventeenth century.[72]

A vast literature surrounds the history of the so-called Hindu-Arabic numerals. Whereas the first epigraphic evidence for zero occurs in India in the Bhojadeva inscriptions at Gwalior, around A.D. 870, much earlier records illustrating the use of place value come from Cambodia (604), Champa (609), and Java (732). Again, a Cambodian inscription of 683 uses the dot or *bindu* to represent zero, whilst an inscription on Banka Island of 686 shows the closed ring. These are no doubt the result of the Hindu influence in South-East Asia, just as in the case of the Po-lo-men (brahman) books transmitted to China during the great period of Buddhist interchange (say 350-1050), or the calendrical texts of eastern Turkestan which derive from Sanskrit, or the earlier dissemination of Indian ideas amongst the Greeks, Iranians, and Chinese in Khotan in the days of Kanishka. Confining ourselves, through the limitations of space, to questions of origin only, we may quote a happy suggestion of Joseph Needham: 'It may be very significant that the older literary Indian references simply use the word 'sunya'—emptiness, just as if they are describing the empty spaces on Chinese counting boards.'[73]

During the medieval period the medical tradition of Caraka and Susruta was continued by the Buddhist Vagbhata (perhaps eighth century) who summarized the eight divisions of medicine in his *Ashtangahridaya Samhita*.[74] Importance was attached to the use of minerals and natural salts in prescriptions, and books of such prescriptions appeared in popular medicine.[75] Medicine and chemistry were closely allied. Alchemy was an integral part of Tantric mysticism. Throughout the Tantric period (c. 700-*c*. 1300) and the ensuing Iatrochemical period (1300-*c*. 1550) the philosophy of mercury, which in the *Rasaratnakara* of the alchemist Nagarjuna and the Saiva tantra *Rasarnava* was concerned essentially with the elixir of life, gradually developed, as exemplified in the *Rasaratna Samucchaya* and many other similar treatises, into the more realistic study of mercurial remedies and the chemical behaviour of the metals.[76] Indian alchemy reached Tibet[77] in the early eighth century with the spread of Buddhism, and is to be found in the great scriptures *Kanjur* and *Tanjur*. In the field of metallurgy remarkable technological competence was attained as early as the fourth and fifth centuries; the casting of the pure-copper Buddha at Sultanganj in Bihar and the welding of wrought-iron shapes to complete the Iron Pillar near Delhi cannot fail to inspire the highest respect.[78]

IV

In astronomy the Muslim tradition of instrumental technology survived in India until the middle of the eighteenth century.[79] The astrolabe, which had been lovingly perfected by generations of Persian and Arab craftsmen and was again executed in fine workmanship by the family of 'Isa b. Allahdad in Lahore[80] in the reign of the Mughal Emperor Jahangir (1605-27), was used by the astronomers in the service of the Maharaja Sawai Jai Singh II (1686-1743) at his observatories in Delhi, Jaipur, Ujjain, Varanasi (Benares), and Mathura.[81] Though Jai Singh's principal astronomer was the Hindu Jagannath he made full use of European and Islamic ideas. In particular his massive masonry quadrants and dials, constructed to attain maximum accuracy, in the absence of the telescope in India, closely follow the precedent set by the Samarkand observatory of Ulugh Beg. On a much simpler level the gnomon, which in Borneo[82] consists of a vertical staff placed in the ground (the cotangent form), appeared with a short horizontal piece inserted near the top of the staff (the tangent form). A shepherd's timestick of the latter type, inscribed in *nagari* script and from Nepal, has been described elsewhere by the present writer.[83]

During the late eighteenth and most of the nineteenth centuries Europeans resident in India and South-East Asia, excited by the new world of natural history around them, began the process of describing and classifying the native flora and fauna. Thus the remarkable realism of the Mughal court painters,[84] especially of Ustad Mansur under the patronage of Jahangir, himself an ardent naturalist, was followed by a still more accurate art form dictated by the requirement of scientific recording. Beautiful drawings by Indian, Chinese, and European artists survive from this period.[85]

India has not failed in this century to produce her Fellows of the Royal Society. We conclude with the three names of Sir C. V. Raman, Sir J. C. Bose, and S. Ramanujan. Raman investigated both experimentally and theoretically the general problem of the molecular scattering of light,[86] which includes an explanation of the colours of the sky and the sea. In the course of these investigations he discovered in 1928 that when a transparent liquid is irradiated by a strong source of light of frequency n the spectrum lines as seen through the spectrometer used to examine the scattered light contain not only the exciting line n but several weaker lines of frequencies $n \pm \Delta n$ on either side. The small values Δn depend not upon n but only upon the nature of the irradiated liquids. This phenomenon has been named the Raman Effect and explained in terms of the quantum theory of Einstein and Planck. Bose crossed the boundaries dividing physics and biology.[87] In a remarkable series of researches during the period 1895-1927, in which the traditional Hindu sensitivity to the living world of plants and animals attained a new understanding, he made a unique contribution to our knowledge of physiological response. In the course of this work he devised delicate apparatus to measure extremely short intervals of time and rates of reaction. Of his high magnification crescograph, which could detect a rate of growth in plants of the order of one millionth of a millimetre per second, he wrote: 'So sensitive is the recorder that it shews a change of growth-rate due to slight increase of illumination by the opening of an additional window.'[88] Bose reached the ultimate in the study of the inertia of mechanical systems prior to the development of the cathode-ray oscillograph and the new methods of electronic engineering. In Srinivasa Ramanujan (1887-1920) we see once again the brahman mathematician inspired by the theory of numbers, leaping intuitively to generalizations whilst the less gifted ponder on the intermediate steps.[89] The 'quiet, meditative child who used to ask questions about the distances of the stars', who later said that the goddess of Namakkal fed him with formulae in his dreams, and who helped to create the beautiful expressions in the Rogers–Ramanujan identities,[90] developed, as G. H. Hardy has recorded, an 'insight into algebraical formulae, transformation of infinite series, and so forth, that was most amazing. On this side most certainly I have never met his equal, and I can compare him only with Euler and Jacobi.'[91]

In this survey emphasis has been laid upon the more recent investigations; the older works, important though some of them, such as those of Colebrooke and Thibaut, still are, have been only briefly mentioned, but any further information relating to them may be found by studying the material listed in the select references at the foot of each page;[92] with the rapid development of researches into the history of science in Asia in the last few years priority of space must be given to the results of these researches, especially those obtained by native scholars. Thus, the pioneer writings of Sudhakara Dvivedi of Varanasi (Benares) on Hindu mathematics have now been supplemented by the publications on ancient astronomy and mathematics to the memory of Sir Asutosh Mookerjee by the University of Calcutta and by the current series of texts being issued by the Department of Astronomy and Mathematics of Lucknow University. Islamic science is represented by Dr. Nizamuddin through the impressive series of publications of the Osmania Bureau of Hyderabad in the Deccan.[93] Two conferences in 1952 and 1961 have helped to clarify ideas,[94] and research papers are now appearing in the newly established *Indian Journal of the History of Science.*[95]A recent attempt by Finnish philologists to read the Indus Valley script using a computer method seems to indicate that the *nakshatras* are of Harappan origin, as also are the later Dravidian names of the five planets related to their colours (e.g. Mars, the 'red star'). Should this be substantiated, it would locate the origin of the *nakshatras*, traditionally associated with the Hindus, within the earlier Indus Valley culture.

It is in Sanskrit that the soul of India lies; the problem is one of interpretation of the *slokas* or stanzas which enshrine the ancient wisdom, a process often rendered more difficult in astronomy and mathematics by the condensed nature of the original presentation. Moreover, many manuscripts have been lost, for example the eighth-century writings *Patiganita* and *Siddhanata Tilaka* of Lalla and the ninth-century *Govindakriti* of Govinda; though there are others being discovered and examined. It is desirable that the significant birch-bark and palm-leaf manuscripts should be dated by the Carbon-14 process.[96]

Ancient Hindu science, though conservative, is strong in classification. This is immediately evident not only in medicine but in the world's earliest scientific grammar, that of Panini, established by the fourth century B.C. It is associative, drawing by intuition from an accumulated background of experience; the direct path of deduction so beloved of the Greeks is alien to it. It lies outside the European tradition. In both ancient and medieval times the native master-pupil relationship had its triumphs and its failures; on the one hand 'schools' would merely perpetuate traditional methods or simply die away, on the other the torch of learning would be kindled anew by the genius of some isolated *guru.*

References

1. Usually II by 5.5 by 2.5 inches.
2. This has been found equivalent to 13-64 grammes in the series of weights.
3. S. Piggott, *op. cit.,* pp. 276-81. An excellent plan and side elevation are drawn to scale on p. 280. See also S. D. Singh, *Ancient Indian Warfare with Special Reference to the Vedic Period,* Leiden, 1965.
4. *Taittiriya Samhita,* iv. 4. 10.
5. This is supported by another list of 27 in the *Kathaka Samhita* (xxxix. 13) and lists of 28 in the *Maitrayani Samhita* (ii. 13, 20) and the *Atharva-Veda* (xix. 7).

6. For Uighur and Tibetan versions see W. Petri, *Indian Journal of History of Science,* Vol. 1, pt. 2, New Delhi, 1966, pp. 83-90.
7. C. P. S, Menon, *Ancient Astronomy and Cosmology,* London, 1931, p. 94.
8. E.g. G. Thibaut. 'On the Surya-Prajnapti', *JASB,* 49 (1880), 107-27, 181-206. B. Datta. 'The Jaina School of Mathematics', *Bulletin of the Calcutta Mathematical Society,* 21 (1929), 114.
9. See G, P. Malalasekera, *The Pali Literature of Ceylon,* R.A.S. London, 1928. M. H. Bode, *The Pali Literature of Burma,* R.A.S. London, 1909. Also issues of the Pali Text Society, Dawson Place, London.
10. W. McGovern, *A Manual of Buddhist Philosophy,* Vol. 1, *Cosmology,* London, 1923.
11. G. S. P. Misra, 'Logical and Scientific Method in Early Buddhist Texts', *Journal R.A.S. London* (1968), Pts. 1-2, pp. 54-64. T. W. Rhys Davids and J. E. Carpenter, *Digha Nikaya,* 3 vols., Pali Text Society, London, 1889-1910, reprinted 1947-1960.
12. Olufsen noted that the Vakhans believe in fire, air, water, and wind, and refers back to the Mazdak sect of the old Iranian religion. We come to the Hindu theory of wind later. O. Olufsen, *Through the Unknown Pamirs,* London, 1904, p. 199. See also P. Ray, *Indian Journal of History of Science,* Vol. 1, No. 1 (1966), 1-14. .
13. R. Shamasastry, *Kautilya's Arthasastra,* 3rd edn., Mysore, 1929. R. P. Kangle, *Kautiliya Arthasastra,* Pt. 1, Bombay, 1960.
14. V. Raghavan, *Yantras or Mechanical Contrivances in Ancient India,* Indian Institute of Culture, Bangalore, 2nd edn., 1956.
15. R. L. Brohier, *Ancient Irrigation Works in Ceylon,* 3 pts., Colombo, 1934-5.
16. See e.g. H. J. J. Winter, 'Muslim Mechanics and Mechanical Appliances', *Endeavour,* 15(1956), 25-8.
17. Bibhutibhushan Datta, *The Science of the Sulba,* University of Calcutta, 1932.
18. Bibhutibhushan Datta, 'Geometry in the Jaina Cosmography', *Quellen und Studien zur Geschichte der Mathematik,* Abt. B. Bd. 1 (1930), 245-54.

19. See e.g. V. Sharma, *Katyayana Sulva Sutra*, Benares, 1928. D. Srinivasachar and V. S. Narasimhachar, *Apastamba Sulva Sutra*, with the commentaries of Kapardisvami, Karavindasvami, and Sundararaja, University of Mysore, 1931.
20. Being the side of a square of area twice that of the original square. Numerically this was usually a surd; thus *dvi-karani* meant $\sqrt{2}$, *tri-karani* meant $\sqrt{3}$ (B. Datta, *The Science of the Sulba*, p. 188).
21. *Baudhayana Sulva Sutra*, i. 59.
22. H. R. Zimmer, *Hindu Medicine* (Hideyo Noguchi Lectures), Johns Hopkins Press, Baltimore, Md., 1948, p. 82.
23. P. J. Deshpande, K.R. Sharma, and G. C. Prasad, 'Contribution of Susruta to the Fundamentals of Orthopaedic Surgery', *Indian Journal of the History of Science*, Vol. 5, No. 1 (1970), 13-35. L. M. Singh, K. K. Thakral, and P. J. Deshpande, 'Susruta's Contributions to the Fundamentals of Surgery', *ibid.*, pp. 36-50. See J. Filliozat, *La Doctrine classique de la medecine indienne; ses origines et ses paralleles grecs*, Imp. Nat., Paris, 1949, Ch. VIII.
24. A fourth humour, blood, was later added and is mentioned in the writings of Caraka and Susruta.
25. As many as ten are listed in the Yoga treatise *Gorakshasataka*.
26. Caraka, vi.5. A. C. Kaviratna and P. S. Kavibhushana, *Charaka-Samhita*, Calcutta, 1890-1911.
27. Susruta, iii. 7. Kaviraj K. L. Bhishagratna, *Susruta*, Calcutta, 1907-18. *Sushruta Samhita*, Chowkhamba Sanskrit Studies. 30. 4 vols.. 1963-4.
28. H. Zimmer, *op. cit.*, p. xlix. J. Filliozat, *op. cit.*, Ch. VIII.
29. See J. Filliozat, *op. cit.*, Ch. IX.
30. Caraka, vi. 9.
31. Caraka, vi. 16.
32. The *Bhela Samhita*, Sanskrit text, University of Calcutta, 1921. See also P. Ray, 'Medicine—as it evolved in Ancient and Mediaeval India', *Indian Journal of History of Science*, Vol. 5. No. 1 (1970), 86-100.

33. A. F. R. Hoernle, *The Bower Manuscript,* Pts. 1, 2, Arch. Survey of India, New Imp. Ser. 22, Calcutta, 1893-1912.
34. The present text of the *Susruta Samhita* probably dates from the fourth century A.D.
35. *Hastyayurveda,* ed. Mahadeva Cimanaji Apte (Anandasrama Sanskrit Ser. No. XXVI), 1894. H. Zimmer, *Spiel an der Elefanten,* Munchen, 1929. Franklin Edgerton, *The Elephant-Lore of the Hindus, the Matanga-Lila of Nilakantha,* New Haven, Conn., 1931. On horse medicine see *Asvavaidyaka* by Jayadatta, and *Asvacitiksita* by Nakula, ed. Umesacandra, Bibliotheca Indica, Calcutta, 1887.
36. S. Gurumurthy, 'Medical Science and Dispensaries in Ancient South India as gleaned from Epigraphy', *Indian Journal of History of Science,* Vol. 5. No. 1 (1970), 76-9.
37. The earlier limit. It may still have been developing as late as A.D. 1000.
38. For further sources, see D. Pingree, 'Sanskrit Astronomical Tables in the United States', *Trans. Amer. Philos. Soc.,* Vol. 58, No. 3 (1968), 77.
39. K. S. Shukla, *The Surya-Siddhanta, with the Commentary of Paramesvara,* Dept. of Mathematics and Astronomy, Lucknow University, 1957.
40. H. J. J. Winter, *Eastern Science,* London, 1952, Ch. III.
41. R. E. M. Wheeler, *Rome beyond the Imperial Frontiers,* London, 1954.
42. V. R. Ramachandra Dikshitar, Translation of the *Silappadikaram,* Oxford, 1939. A. Danielou, Translation of the *Silappadikaram,* New York, 1965. P. T. Srinivas Iyengar, *History of the Tamils from the Earliest Times to 600* A.D., Madras, 1929.
43. W. W. Tarn, *The Creeks in Bactria and India,* Cambridge, 1951. J. Marshall, *Taxila,* 3 vols., Cambridge, 1951. R. E. M. Wheeler, 'Roman Contact with India, Pakistan and Afghanistan', in *Aspects of Archaeology, Essays Presented to O. G. S. Crawford,* London, 1951.
44. S. H. Taqizadeh, *Bull. School of Oriental Studies,* 9 (1939), 133-9. D. Pingree, 'Astronomy and Astrology in India and Iran', *Isis,* Vol. 54, No. 2 (1963), 229-46.

45. O. Neugebauer, *The Astronomical Treatise P. Ryl. 27*, Kgl. Danske Vidensk. Selsk. hist.-filol. Meddelelser, XXXII, 2. 1949. Knudtzen-Neugebauer, 'Zwei astronomische Texte', *Bull, de la Soc. Royale de Lettres de Lund* (1946-7), pp. 77-8.

46. O. Neugebauer, 'Tamil Astronomy', *Osiris*, 10 (1952), 252-76; *The Exact Sciences, in Antiquity*, Copenhagen, 1951, Ch. VI. See also: *Arch. Internat. d'Hist. des Sciences*, No. 31, Paris, 1955, pp. 166-73.

47. O. Neugebauer, 'The Transmission of Planetary Theories in Ancient and Mediaeval Astronomy', *Scripta Mathematica*, Yeshiva University, New York, 1955, p. 7.

48. O. Neugebauer, *op. cit.*, p. 8 and Appendix.

49. Revd. E. Burgess, Translation of the *Surya-Siddhanta*, ed. P. Gangooly and P. Sengupta, University of Calcutta, 1935.

50. O. Schmidt, 'On the Computation of the Ahargana', *Centaurus*, z (1952), 140-80.

51. Ch. II, vv. I-II. Revd. E. Burgess, *op. cit.*, p. 53.

52. K. S. Shukla, *Maha-Bhaskariya* (Bhaskara I and his works Pt. 2), Department of Mathematics and Astronomy, Lucknow University, 1960.

53. Sudhakara Dvivedi, *Sisya-dhi-urddhida of Lalla*,Benares, 1886.

54. Id., *Maha-Siddhanta of Aryabhata* II, *Benares*, 1910.

55. Babuaji Misra, *The Siddhanta-Sekhara of Sripati, with the Commentary of Makkibhatta* (A.D. 1377), University of Calcutta, Pt. I, 1932, Pt. 2, 1947. H.R. Kapadia *Ganitatilaka by Sripati, with the Commentary of Simhatilaka Suri (c.* A.D. *1275*)(Gaekwad's Oriental Ser. 78), Oriental Institute, Baroda, 1937.

56. For source materials see S.N. Sen, *A Bibliography of Sanskrit Works in Astronomy and Mathematics*, National Institute of Sciences of India, Calcutta, 1966.

57. P.C. Sengupta, 'The Aryabhatiyam', *Journ. Dept. of Letters, University of Calcutta*, 16 (1927). W.E. Clark, *The Aryabhatiyam of Aryabhata*, Chicago, 1930. K.S. Sastri, *Aryabhatiya of Aryabhata I, with the Commentary Mahabhasya of Nilakantha* (A.D. 1500) 3 parts, Trivandrum, 1930-6.

58. Islamic astronomers were indebted to Greek and Hindu mathematics and in some respects improved on both. See e.g. Al-Biruni, *Al-Qaniln Al-Mas'udi,* 3 vols., Osmania Publications Bureau, Hyderabad-Deccan, 1954-6. E. S. Kennedy and Ahmad Muruwwa, 'Biruni on the Solar Equation', *Journ. Near Eastern Studies,* Vol. 17, No. 2 (1958), H. J. J. Winter, 'Formative Influences in Islamic Science', *Arch. Int. d'Histoire des Sciences,* 23-4 (1953), 171-92.
59. A Caratheodory, *Traite du quadilatere,* Constantinople, 1891.
60. Sudhakara Dvivedi, *Brahmasphuta Siddhanta and Dhyana-Grahopadesadhyaya by Brahmagupta,* reprint from *The Pandit,* N.S. 24 (1902), Sanskrit text.
61. P. C. Sengupta, *The Khandakhadyaka, with the Commentary of Prthudaka,* 2 vols., University of Calcutta, 1934, 1941. 'A short treatise in astronomy which is as pleasant as food prepared with sugar-candy.'
62. This formula, though arising from the general treatment in the *Brahmasphuta-Siddhanta,* is first explicitly stated in the *Ganitasara Samgraha* of Mahavira (A.D. 850).
63. H. J. J. Winter, *Eastern Science,* London, 1952, p. 47.
64. The so-called Sindhind has not been precisely identified.
65. See the *Patiganita of Sridharacarya,* ed. K. S. Shukla, Lucknow University, 1959, pp. vi, xv.
66. G. R. Kaye, *The Bakhshali Manuscript,* Archaeological Survey of India, New Imperial Series, Vol. 43, Pts. 1, 2, Calcutta, 1927; Pt. 3, Delhi, 1933.
67. H. T. Colebrooke, *Algebra, with Arithmetic and Mensuration from the Sanscrit of Brahmagupta and Bhascara,* London, 1817, p. 156.
68. See H. J. J. Winter, *Eastern Science,* Ch. III. D. Apte, *Bijaganita of Bhaskara II, with the Commentary of Navankura of Krsna* (c. A.D. 1600), Anandasrama Sanskrit Series, Poona, 1930. H. C. Banerji, *Lilavati of Bhaskara II,* Calcutta, 1893. G. P. Dvivedi, *Siddhanta-Siromani by Bhaskara II,* Sanskrit and Hindi, Vols. 1 and 2, Lucknow, 1911, 1926. M., Jha, *Siddhanta-Siromani by Bhaskara II with the Commentaries of Vasanavartika of Nrsimha* (1621) *and Marici of Munisvara* (1635), Vol. 1, Benares, 1917.

H. T. Colebrooke, *op. cit.*, pp. 172-8.

69. H. J. J. Winter and A. Mirza, *Journ. Asiatic Society, Bengal, Science*, Vol. 18, No. 1 (1952), 1-10.

70. See Euclid, *Elements*, ed. I. Todhunter, intro. Sir T. L. Heath, Everyman Reprint, London, 1955, p. 266. J. Needham. *Science and Civilization in China*, Vol. 3, Cambridge, 1959, p. 22. H. T. Colebrooke, *op. cit.*, p. 222.

71. Narayana Pandita, *Ganita Kaumudi*, Princess of Wales Bhavana Texts, Government Sanskrit College, Benares, Vol. 1, 1936; Vol. 2, 1942.

72. De La Loubere. *Description du royaume de Siam*, Vol. 2, pp. 235-88; *Le Probleme des quarres magiques selon les Indiens*, Amsterdam, 1714.

73. J. Needham, *op. cit.*, Vol. 3, p. 11 n.

74. L. Hilgenberg and W. Kirfel, *Vagbhata's Astangahrdayasamhita, ein altindisches Lehrbuch der Heilkunde*, Leiden, 1937-40. K. Vogel, *Vagbhata's Astangahrdaya Samhita. The First Five Chapters of its Tibetan Version*, Berlin, 1965. *But* see also A. B. Keith, *A History of Sanskrit Literature*, Oxford, 1920, p. 510.

75. See e.g. Elizabeth Sharpe, *An Eight-Hundred Year Old Book of Indian Medicine and Formulas, from Old Hindi*, London, 1937.

76. Acharya Prafulla Chandra Ray, *History of Chemistry in Ancient and Mediaeval India*, now edited by P. Ray, Indian Chemical Society, Calcutta, 1956.

77. Ray, op. cit., Tibetan texts, pp. 449-56.

78. See further, *Indian Journal of History of Science*, Vol. 5, No. 2 (1970), Sections X-XII.

79. H. J. J. Winter, 'The Muslim Tradition in Astronomy', *Endeavour* 10 (1951).

80. J. Frank and M. Meyerhof, *Ein Astrolab aus dem indisches Mogulreiche*, Heidelberg, 1925.

81. G. R. Kaye, *The Astronomical Observatories of Jai Singh*, Archaeol, Survey of India, New Imperial Series, Vol. 40, Calcutta, 1918.

82. J. Needham, *op. cit.*, Vol. 3, p. 286.

83. H. J. J. Winter, *Physis*, Vol. 4, Pt. 4 (1964), 377-84.

84. S. M. Hasan, 'The Mughal School of Zoological Portraiture',

Arts and Letters, Journal Roy, India, Pakistan and Ceylon Society, Vol. 37, No. 1 (1963), 3-13.

85. Mildred Archer, *Natural History Drawings in the India Office Library,* H.M.S.O., London, 1962.
86. C. V. Raman, *The Molecular Diffraction of Light,* University of Calcutta, 1922. Raman and Krishnan, *Proc. Roy. Soc.* 122 (1929), 23.
87. P. Geddes, *The Life and Work of Sir Jagadis C. Bose,* London, 1927. Sir Jagadis Chunder Bose, *Collected Physical Papers,* Trans. Bose Inst. Calcutta, and Longmans Green, London, 1927.
88. *Ibid.*, p. 350.
89. G. H. Hardy, P. V. Seshu Aiyar, and B. M. Wilson, *The Collected Papers of Srinivasa Ramanujan,* Cambridge, 1927.
90. Proof of certain identities in combinatory analýsis', *Proc. Comb. Philos. Society,* 19 (1919).
91. *The Collected Papers of Ramanujan,* p. xxxv.
92. Or W. E. Clark in the *Legacy of India,* pp. 335-68.
93. See current catalogue of the Arabic publications of the Dairat ul-Ma'arif 'il- Osmania, Hyderabad, Deccan.
94. See e.g. 'Symposium on the History of Sciences in South Asia', *Proc. Nat. Inst. of Sciences of India,* Vol. 18, No. 4 (1952), 323-62.
95. One volume has appeared annually since 1966. *A Concise History of Science in India* by D. M. Bose, S. N. Sen, and B. V. Subbarayappa has just been published by the Indian National Sciences Academy (Calcutta, 1973).
96. W. F. Libby, *Radiocarbon Dating,* University of Chicago Press, 2nd edn., 1955.

2
Hindu Calendar

The early religion of the Hindus reveals an intimate knowledge of the times and seasons, and there was from the earliest times, to regulate their rites and ceremonies, a calendar setting forth the order in which they should be observed. This calendar, in the earlier periods, was of an imperfect character, which led to methods afterwards adopted for its improvement, generally with a view to its adaptation to religious rather than to secular uses.

Now, among all the nations, the fundamental periods of time, the day, the month, the year, are the same, the variations occurring in them being principally in the arrangement of the days to form months and years; in the sub-divisions of the day; in the times to be reckoned as the commencement of the day; whether at mid-night, sun-rise, or noon; in the sub-divisions of the year into months, differing from each other as to the number of days of each; in the various kinds of months to form the year, and the like. There have been in all nations certain difficulties experienced as to the time when the year should be reckoned to begin, and in the consequent arrangement of the months and seasons, so that they should recur at regular intervals.

The present Indian system has been thus described by Whitney:[1] "In the ordinary reckoning of time, these elements are variously combined. Throughout Southern India the year and month made use of are solar, and the day civil; the beginning of each month and year being counted, in practice, from the sun-rise nearest to the moment of their actual commencement. In all Northern India the year is luni-solar, the month is lunar and is divided into both lunar and civil days; the year is composed of a variable number of months, either twelve or thirteen, beginning always with the lunar month, of which the commencement next precedes the true commencement of the sidereal year. But underneath this division, the division of the actual sidereal year into twelve solar months is likewise kept up, and to maintain the concurrence of the civil and lunar days, and the lunar and solar months, is a process of great complexity, into the detail of which we need not enter here."

These complications are, however, of later growth. The four ways of reckoning time, the Savana, the Candra, the Naksatra and the Saura, are not all referred to in the early works, and even in the later days all these measures of time were not fully utilised. The information about the calendar used in the Vedic times is gathered from stray references in the Vedic and classical literature of India. Tilak remarks, "There are several sacrificial hymns in the Rgveda, which show that the sacrificial ceremonies must have been considerably developed; and as no sacrificial system could be developed without the knowledge of months, seasons, and the year, it will not be too much to presume that in the Vedic times there must have existed a calendar to regulate the sacrifices. It is difficult to determine the exact nature of this calendar, but a study of the sacrificial literature would show that the phases of the moon, the changes in the seasons, and the southern and northern courses of the sun were the principal landmarks in the measurement of time in those early days. What is still more interesting, however, is that the leading features in the early sacrifices are the same as those in the year."[2]

The ancient Rsis, therefore, prepared their calendar mainly for sacrificial purposes, and the performance of various sacrifices facilitated the maintenance of the calendar. When the course of sacrifices was completed, it was found that the year also had run its course, and the sacrifice and the year, therefore, became synonymous terms. There are passages in the Brahmanas where Samvatsara and Yajna are used as convertible terms.[3]

The difficulties experienced by the Hindus in adjusting their calendar occasioned repeated changes of their system. "At one period the motion of the moon was taken as its foundation, and the lunar month was formed to agree with the phases of the moon. Then a change took place, and a solar month was formed, constituted so as to be reckoned by the time the sun, in its progress, remained in each sign of the solar Zodiac. Another change followed, efforts being made to reconcile the two previous systems, in which each kind of month preserved its original character, the solar month being reckoned in ordinary civil days, and the lunar months measured by *tithis* or lunar days, each being one-thirtieth part of a synodic period, the time elapsing between two conjunctions of the sun and the moon The result of these efforts was the formation of the luni-solar year, either in civil days or in *tithis*."[4]

The ancient Hindus were struck with the daily appearance of the moon in the heavens and observed that the moon had a regular course among the stars. The moon moved from a star and came back to the same in 27 solar days. Hence they fixed the path of the moon in the heavens, and, because they found that the moon finished this course in 27 solar days, they divided the path in so many asterisms.

Moreover, the ancient Hindus found that the moon totally disappeared one night and again became full and round another night. They called these phases new and full moon and further observed that from one new moon to another or from one full moon to another the sun rises thirty times.[5] Hence one lunar month became equivalent to thirty days. The ancient Hindus further observed that the star that rose or set at sun-rise one day, would not do so after the lapse of several days.[6] They concluded that the sun like the moon moved among the stars in the heavens and that the sun took 12 months to complete this course. Thus according to this calculation a month consists of 30 days and a year contains 12 months.[7]

The Hindus from continued observations invented from their imaginative brain a table or arrangement beginning from the infinite time and ending with the minutest divisions of day. In the first chapter of *Manusamhita*, we get the following arrangement:

"The sun causes the division of day and night, which are of two sorts, those of men, and those of the gods; the day for the work of all creatures in their several employments, the night for their slumber. A month is a day and night of the Pitris, and it is divided into two parts; the bright half is their day for work, and the dark half their night for sleep. A year is a day and night for the gods, and that also is divided into two parts; the day is when the sun moves towards the North; the night when it moves towards the South. Learn now the duration of the day and night of Brahma, with that of the ages, respectively and in order. Four thousand years of the gods they call Krita (or Satya) age; and its limits at the beginning and at the end are in like manner as many hundreds. In the three successive ages, together with their limits, at the beginning and end of them are thousands and hundreds diminished by one. This aggregate of four ages, amounting to twelve thousand divine ages added together, must be considered as a day of Brahma. His night has also the same duration. The above mentioned days of the gods, or twelve thousand of their years multiplied by seventy-one, form what is named here a Manvantara. There are alternate creations and destructions of worlds through innumerable Manvantaras. The Supreme Being performs this again and again."

A similar arrangement is given in astronomical works of later times, but with the ages reckoned in years of mortals. For comparison, the following extract is taken from the Surya Siddhanta[8]:

"A solar year consists of twelve solar months; and this is called a day of the gods.[9] An Ahoratra (day and night) of the gods and that of Asuras (demons) are mutually at the reverse of each other (viz., a day of the gods is the night of the demons, and, conversely, a night of the gods is a day of the demons). Sixty Ahoratras multiplied by six make a year of the gods and of the demons. The time containing twelve thousand years of the gods is called a Caturyuga (the aggregate of the four yugas, Krta, Treta, Dvapara and Kali). These four yugas, including their Sandhya and Sandhyamsa,[10] contain 4,320,000 years. The tenth part of 4,320,000, the number of years in a Great Yuga, multiplied by 4,3,2,1, respectively, make up the years of each of the four yugas, Krita, Treta, Dvapara and Kali, including their sixth part, which is collectively the number of years of Sandhya and Sandhyamsa (the periods at the commencement and expiration of each yuga). Seventy-one Great Yugas (containing 306,720,000 solar years) constitute a Manvantara (a period from the beginning of a Manu to its end), and at the end of it, 1,728,000, the whole number of solar years of the Krita, is called its Sandhi; and it is the time when a universal deluge occurs. Fourteen such Manus, with their Sandhis, constitute a Kalpa, at the beginning of which is the fifteenth Sandhi, which contains as many years as a Krita does. Thus, a thousand of the Great Yugas make a Kalpa, a period which destroys the whole world. It is a day of the god

Brahma, and his night is equal to his day. The age of Brahma consists of a hundred years, according to the enumeration of day and night. One-half of his age has elapsed, and this present Kalpa is the first in the remaining half of his age."

Mention of Kalpa, as a measure of time, though found in the Puranas and Samhitas of a later date, is nowhere to be found in the Vedas. The peculiar form in which the construction of the Kalpa is expressed attracted much attention in the western countries more than two centuries ago, and various theories were put forward to account for it. For the purpose of examining this construction, following backwards the order in which the Kalpa has been formed, we get:

1 Kalpa = 14 Manvantaras +1 Krtia

The Manvantara = 71 Mahayugas + 1 Krita

1 Mahayuga = 10×432,000 years

Krita = 4×432,000 years

Therefore, the Manvantara

= 710×432,000 + 4×432,000

= 714×432,000

Hence, the Kalpa = (14×714 + 4) × 432,000 years

= 4,320,000,000

It will be seen that this number consists of two factors, 14 × 714+4, which has the form mn + r= 10,000, and the co-efficient 432,000. In this connection Brennand remarks: "The form of the number shows that its inventors had a special design in view in its construction, i.e., to multiply the Kali period with the significant figure 432 unchanged. If they had no other design, there would have been no reason why they should have deviated from the rule laid down in the Institutes of Manu, which only required that they should multiply the divine age by a thousand. If they had merely wished to multiply 432,000 by 10,000, they would not have taken the trouble of putting the operation into such a singular form. It is clear that they did not wish to alter the factors already existing, in the Kali Yuga, namely, 60×60×60×2, and that they specially wished to multiply by 10,000, so that their system would still be in conformity with that which was established in the Institutes of Manu and in the Vedas. Now there are a great many ways in which they might have multiplied by 10,000, and the fact that they selected the special form (14×714+4) shows a design. The number is one out of the set mn+r = 10,000.[11] If we take m to be any number less than 100 which is not one of the eleven factors of $5^4 \times 2^4$ (each of which would divide 10,000 without a remainder), it would find by division a number which would have the form mn + r, and there would be 89 such cases, thus,

10,000 = 3 × 3,333 + 1 = 6×16,666+4 = 7 × 1,428 + 4

= 9 × 1,111 + 1

= 11 × 909 + 1 = 12 × 833+4 = 13×769+3 = 14×714+4.

And so on, we might go through the whole of the 89 cases. Out of all these cases, it is incredible that the particular form 14×714+4 should have been selected by chance.[12]

It is, therefore, almost certain, that the astronomers who invented the Kalpa had made the discovery that in 714 years the Solstice had retrograded approximately 10°. Then the Solstice would have gone back 14° in 1,000 years or 140° in 10,000 years. In 140° there are 504,000 seconds, which, divided by 10,000 gives the precession equal to 50.4." This coincidence could not have happened by chance where there were so many adverse cases. Hence we may conclude that the construction of the Kalpa was designed to include a correct estimate of precession.

Next we come to the Mahayuga. There are references to the Mahayuga and the various yugas in the Vedas, the

Samhitas and the Brahmanas. We have already seen in the extracts from *Manu Samhita* mentioned before that a Mahayuga, which is the aggregate of the four yugas amounts to twelve thousand years of the gods; and one day (including night) of the gods is one year of the mortals, i.e., 360 Saura days. The ancient Saura day is the variable time which the sun takes in its motion over each degree of the Ecliptic, the aggregate being the same as the number of parts into which the circle of the Ecliptic is divided; and from this, the apparent sidereal revolution of the sun, or the sidereal year is 360 Saura days. But the Hindu astronomers also reckoned the sidereal year in mean solar time to be 365 days 6 hours 12 mins. 36 secs., according to *Pulisa Siddhanta*, or as a mixed number to be $= 365\frac{207}{800}$ mean solar days. The *Surya Siddhanta*[13] adds a fraction of a second more to the length of the year, the Brahma Siddhanta makes the year less by about 27 seconds, and in the *Arya Siddhanta* it is made less by about six seconds.

Therefore, the Mahayuga = 12,000 × 360 Saura years = 4,320,000 Saura years, one year of the gods being equal to 360 Saura years,= $360 \times 365\frac{207}{800}$ mean solar days. The Mahayuga is, therefore, equal to 1,577,917,800 mean solar days according to the Pulisa Siddhanta. The *Surya Siddhanta* makes it 28 days more, i.e., 1,577,917,828[2] mean solar days and the *Brahma Siddhanta* makes it 1350 days less. The Mahayuga again was further subdivided and made up of the Krta, Treta, Dvapara and Kali yugas, in the following manner:

The Krita = four times the Kali and four-tenths of 4,320,000 = 1,728,000

The Treta = thrice the Kali and three-tenths of 4,320,000 = 1,296,000

The Dvapara = twice the Kali and two-tenths of 4,320,000= 864,000

The Kali = one tenth of 4,320,000 = 432,000

Thus the total is = 4,320,000 years

At each of these commencing epochs Hindu astronomers considered that the moving celestial bodies were in conjunction.

A difference of opinion existed among the more ancient astronomers as to whether their calculations ought to begin from the beginning of Brahman's life or the beginning of a Kalpa; and it is suggested in the Surya Siddhanta that the end of the Krita yuga is a convenient epoch, from which to compute easily the terrestrial days and to find the mean places of the planets. It says:[14]

"At the end of the Krita yuga the mean places of all the planets, except their nodes and apogees, coincide with each other in the first point of stellar Aries. At the same instant the place of the moon's apogee is nine signs, her ascending node is six signs, and the places of the other slow-moving apogees and nodes, whose revolutions are mentioned before, are not without degrees (i.e., they contain some signs and also degrees)."

But it will be seen that any one of the above epochs might be used for the purpose of computing the mean places of each; and as the Kali, the smallest period of all, was just as useful as the others for their purpose, it alone was generally used. So that this epoch is one of great importance in considering problems affecting Hindu astronomy as well as questions relating to their civil time.[15]

We have already said that yuga, as a measure of time, is mentioned several times in the Vedic literature.[16] It is also stated that three yugas have already elapsed.[17] In the *Taittiriya Samhita* these three are mentioned by names, viz., Krita, Treta and Dvapara.[18] In the *Vajasaneyi Samhita*, also these names are given.[19] Then in the Brahmanas all the four yugas are mentioned by names,. viz, Krita, Treta, Dvapara and Kali.[20] It may be noted here that in the Vedanga Jyotisa, as well as in the Vedas, a cycle of five (or six years) is used which is also called a yuga.[21] These five were called samvatsara, parivatsara, idavatsara, idvatsara, vatsara[22] (or anuvatsara).[23] Though these five years for a cycle are mentioned in most places as samvatsara, parivatsara, idavatsara, anuvatsara and idvatsara, yet sometimes only four years are mentioned by names as in Tait. Br., i, 4, 10. Sometimes six years also are given for a cycle and the name of *iduvatsara* is added to the list.[24] But the cycle of five years was more current not only at the time of the Vedas, but also after the vedic period and even at the time of the Vedanga Jyotisa.[25] The importance of the cycle of five or six years is this:—a lunar month consisted of 30 days, and 360 days were regarded to constitute a year: this was about 5½ days too short. This required an intercalation to be admitted in the quinquennial period by doubling one month. There were, moreover, the two cycles of Jupiter. One was Brihaspati's (or Jupiter's) year consisting of nearly 12 Saura years, the other the cycle consisting of 60 years. Each of the sixty years was called Brihaspati Mana, or Madhyama, his mean motion through one sign These years had each a separate name. The cycle of Brihaspati of 12 years, as described by Parasara, quoted by Varahamihira, is thus explained:

"The name of the year is determined from the Naksatra in which Brihaspati rises and sets heliacally, and they follow in the order of the lunar months. The years beginning with Kartika commence with the Naksatra Krittika, and to each there appertain two Naksatras, except the 5th, 11th and 12th years, to each of which appertain three Naksatras."

There was a difference of opinion amongst the Hindu astronomers regarding the naming of these years. The names and order of the twelve Brhaspati years were not the same as those of the cycle of 60. According to Sasipura and others, the

Naksatra in which the Jupiter rises, gives the name to the year. Kasyapa says that the name of the Samvatsara yuga and the years of the cycle of 60 are determined by the Naksatra in which he rises. Garga gives the same account. Some make the cycle to begin on the first day of the month of Caitra, etc., whatever may be the Naksatra in which Brhaspati stands. According to Parasara, who mentions also the character distinguished as good or bad with the names and order of the corresponding Naksatras, they are as follows, with the presiding deities:

Years	*Naksatras*	*Deites*	*Character*
Kartika	Krittika, Rohini	Vishnu	Bad
Agrahayana or Margasirsa	Mrgasiras, Ardra	Surya	Bad
Pausa	Punarvasu, Pusya	Indra	Good
Magha	Aslesa, Magha	Agni	Bad
Phalguna	Purva-phalguni, Uttara-phalguni, Hasti	Tvastri	Neutral
Caitra	Citra, Svati	Ahivradhna	Good
Vaisakha	Visakha, Anuradha	Pitris	Bad
Jyaistha	Jyestha, Mula	Vishva	Bad
Asadha	Purvahsada, Uttarahsada	Soma	Good
Sravana	Sravana, Dhanistha	Indragni	Good
Bhadra	Satabhisaj, Purva-bhadrapada, Uttara-bhadrapada	Asvini	Good
Asvina	Revati, Asvini, Bharani		

The commentator on Parasara says: "It is in the *Soma Siddhanta* that the presiding deities are thus stated."

In the cycle of 60 years are contained five cycles of twelve years, which five cycles or yugas have already been named—

Samvatsara, over which presides Agni
Parivatsara, over which presides Aditya
Idavatsara, over which presides Candramas
Anuvatsara, over which presides Vayu[26] or Brahman
Idvatsara, over which presides Siva.

Parasara says: "the first year of the cycle of sixty, named Prabhava, begins when, in the month of Magha, Brhaspati rises in the first degree of the Naksatra Dhanistha, because when Brhaspati rises in 9 signs 23° 20', Surya (the sun) must be 10 signs 6° 12'." Parasara gives the names of all the sixty years of Brhaspati, beginning with Prabhava as the first.[27]

Brennand remarks that "this cycle of 60 years was brought into India by the immigrant tribes who came to settle in India from Central Asia, and was afterwards" known as the cycle of Brhaspati. It is a combination of two cycles, a cycle of five years from the Vedas or Jyotisa Vedanga, and the sidereal period of the planet Brhaspati, which was at first reckoned to be 12 years, but was afterwards found by the Hindus to be 11.860962 years. According to Laplace, the mean sidereal period is 11.862 Julian years, or & .138 of a year short of 12 years, an error of about 8¼ months in 60 years, and would, therefore, require periodic correction."[28]

It will be interesting to note here that the Chinese History and the Annals of the Chinese emperors were written by reference to cycles of 60 years. Such a period of time was, moreover, in common use in Chaldea, under the name of Sosos, as mentioned by Berosus. It is stated by several writers, both Persian and Greek, that, besides the Sosos of 60 years, the Chaldeans had in use several other cycles, one of 600 years, called the Neros, another of 3,600 years. They had also a period called the Saros, consisting of 323 complete lunations in 19 years, after the expiry of which period, the new and full moons fall on the same days of the year, i.e., recur as in the previous period.

Now we proceed to discuss the character of a year. We have already said that the sacrifice and the year seem to have early become synonymous terms, and that there are many passages in the Brahmanas and Samhitas, where Samvatsara and Yajna are declared to be convertible terms. It is believed that the Vedic Rishis kept up their calendar by performing the corresponding round of sacrifices on the sacred fire that constantly burnt in their houses, like the fire of the Parsi priest in modern times. The etymological meaning of the word rtvij (rtu+yaj = officiating priest at *season sacrifice*) shows that even in the oldest days there existed a certain correspondence between the sacrifices and the seasons, and what is true of the seasons is true of the year, which, according to one derivation of Samvatsara, is nothing but a period where seasons dwell, or which is a cycle of seasons.[29] We have now to examine the principal parts of the year i.e., the sacrifice. The *Rigvedic* Rishis said, "Who knows that there is one wheel (cạkra), twelve circumferences, three foci and 360 spokes?"[30] The wheel is the Samvatsara, the twelve circumferences are the twelve months, the three foci are the three four-monthly satras (sacrifices) and the 360 spokes are the 360 days in one year. The Savana or the civil day appears to have been, as its etymology shows, selected[31] in these cases as the natural unit of time; 30 such days made a month, and 12 such months or 360 savana days made a year.[32] But a year of 360 days was about 6 days too short. For this reason, the sun and the moon would not return at the end of a year to the same star as was their starting point at the beginning of the year. In five years the number of days was about 30 days too short. Hence unless in a cycle of five years one month is added to the twelve months, the harmony between the months and the year and therefore between the year and the seasons is lost. Now a month of thirty civil or Savana days cannot correspond with a lunar synodical month, and the Brahmavadins had therefore to omit a day in some of the savana months to secure the concurrence of the civil and the lunar months.[33] The year of 360 Savana days was thus practically reduced to a lunar year of 354 civil days or 360 tithis. But a further correction was necessary to adjust the lunar with the solar reckoning of time. The commencement of the cycle of seasons was, therefore, the only means to correct the calendar. The devise of the intercalary days or month was for that purpose very early hit upon by the ancient Hindus. The Vajasaneyi and Taittiriya Samhitas, the Brahmanas and the *Rigveda* mention the intercalary month.[34] The *Rigveda* says, "He who is dhrtavrata (observant of penances) knows the twelve months with their respective merits and also the thirteenth month that is intercalated."[35] It is, therefore, an undisputed fact that in the old Vedic days means were devised and adopted to secure the correspondence of the lunar with the solar year. The ancient Hindus found that though 30 tithis or lunar days made one month, 12 lunar months or 360 lunar days did not make one year. The sun takes 366 Savana days to complete one revolution. Hence 366 civil or Savana days were reckoned to constitute a year. We have already said that the twelve lunar months were reduced to 354 civil or Savana days. Therefore twelve (366 - 354=12) days were required to balance the lunar with the solar year. They were in fact the supplementary days or special days. The sacrificial literature of India still preserve the memory of these days by ordaining that a person wishing to perform yearly sacrifice should devote twelve days (dvadasaha) before its commencement to the preparatory rites.

Thus we find that a complete lunation is measured by 30 lunar days, some of which, of course, must in alternate months be sunk, to make the dates agree with the Nychthemera, for which purpose the sixty-second day appears to be deducted, and thus the cycle of five years consists of 1860 lunar days or 1830 Nychthemera, subject to further correction. As twelve days fall short in one year, a month will have to be added after 32 solar months, or after 2 years and 8 months, and again, two months after 5 years and 4 months. From this it is obvious that a cycle of five years was too short for making the intercalation; a much longer cyclic period was necessary, so that an exact number of lunar months shall coincide with an exact number of solar months, and so that only a small fraction of a year or no fraction at all shall remain. The rule in the Vedas for subtracting the sixty-second day is not quite so correct as that of Bhaskara, who says that the subtracted day occurs in $64\frac{1}{11}$ lunar days (tithis).[36] In this connection Colebrooke says that "the progress of these corrections may be traced from the cycle of five to that of sixty lunar years (which is noticed in many popular treatises on the calendar, and in the commentary of the Jyotisa) and thence to one of 60 years of Brihaspati (Jupiter), and finally to the greater astronomical period of 12,000 years of the gods, and a hundred years of Brahma."

The *Taittiriya Samhita* sums up the different constituents of a year thus: "There are six seasons, twelve months and also thirteen months in a year, fifteen nights in a half-month, twenty-four half-months[37] in a month[38], and three hundred and sixty nights in a year."[39]

One question now arises whether the solar year, with reference to which the above corrections were made, was tropical or sidereal. It is true that the primary object of the calendar was to ascertain the exact time of the seasons. But the changes in the seasons consequent upon the precession of the equinoxes are so exceedingly minute as to become appreciable only after hundreds of years. Under the primeval system, therefore, the year would naturally be said to be complete when the sun returned to the same fixed star. Prof. Whitney has pointed out that the same system is followed in the *Surya Siddhanta*, though the motion of the equinoxes was then discovered.[40] Therefore, the solar year mentioned in the Vedic works was sidereal and not tropical.

The difference between the sidereal and the tropical year is 20.4 minutes, which causes the seasons to fall back nearly one lunar month in about every two thousand years, if the sidereal solar year be taken as the standard of measurement. When these changes and corrections were noted for the first time, they must have created a great deal of surprise, and it was not till after one or two adjustments on this account were made that their true reason, the motion of the equinoxes, could have been discovered.

The next division of time was the month. The Vedic month was lunar and was reckoned from one full moon to another full moon and, later, sometimes, from one new moon to another new moon.[41] We have already said that the word मास is derived from चन्द्रमस् and a month at first meant a lunar month only. The etymology of the word पूर्णमास shows that a month has been full or complete. The solar months came into use with the invention of the signs of the Zodiac and we know the signs Mesa (Aries) etc. were not known at the time of the earlier Vedic literature. We know that the Vedic literature—Samhita and Brahmana—falls into two classes, one part written at an earlier period and the other at a later period. Even the different chapters of the same book come under this classification. For example, the Satapatha Brahmana contains fourteen Kandas, of which ten Kandas containing sixty chapters were written at an earlier period and the remaining four Kandas containing thirty four chapters were of later origin. The nomenclature given to the twelve months in the earlier period is a peculiar one and differs entirely from the later nomenclature. Let us first discuss the earlier nomenclature. A detailed description of this nomenclature is given in the *Satapatha Brahmana*.[42] It is also given in the *Taittiriya Samhita* and the *Vajasaneyi Samhita* in detail.[43] Madhu and Madhava are the spring months when plants sprout and trees are brought to ripeness; Sukra and Suci are the summer months, when the sun burns fiercest and clearest: Sukra means clear and Suci means bright; Nabhas and Nabhasya are the months of the rainy season: Nabhas means mist or cloud; Isa and Urja are the autumn months, because in autumn food (urja) and juice (plants) ripen; Sahas and Sahasya are the winter months, because the winter by force (sahas) brings the creatures into his power; Tapas and Tapasya are the months of the dewy season, because during them it freezes most severely.

At the time we are speaking of, the intercalary month was called by three names, viz., samsarpa, malimluca and amhasaspati.[44] The Aitareya Brahmana remarks that practically there is no thirteenth month.[45] It may be pointed out here that under the present system of nomenclature the intercalary month has got no separate name; when there is an intercalary month in any year, that month is named after the previous month with the prefix *adhika*; for example, if an intercalary month comes after Vaisakha, it is called Adhika Vaisakha; the year 1850 Saka Samvat, the current year in the Upper India, has got an intercalary month after Sravana and this is called Adhika Sravana. The intercalary month was generally introduced after 32 or 33 months; but according to the Vedanga Jyotisa the intercalary month should come after thirty months. Three denominations of the intercalary month, viz., samsarpa, malimluca, and amhaspati (or amhasaspati) were further distinguished. "The lunar month passed over by the sun is called Malimluca" (*Vyasa Samhita*); it is further explained thus: "If the sun passes over one sign in two months, the earlier month is called Malimluca, and the latter, the pure (regular) month" (*Maitreya Sutra*). "The intercalary month without a Samkranti day is called Samsarpa and the intercalary month which contains two Samkranti days is called Amhaspati" (*Narada Samhita*). Samkranti is the time when the sun passes from one sign to another. Hence in one intercalary month the sun does not pass on to another sign, in the other the sun passes twice.

Another system of nomenclature for the months, half-months and seasons is given in the Taittiriya Brahmana.[46] The months are as follows: Aruna, Arunarajas, Pundarika, Visvajit, Abhijit, Ardra Pinvamana, Annavan, Rasavan, Iravan, Sarausadha and Sambhara; the thirteenth month was named Mahasvan.[47] The half-months were: Pavitra, Pavayisyan, Puta, Medhya, Yasas, Yasasvan, Ayu, Amrta, Jiva, Jivisyan, Svarga, Loka, Sahasvan, Sahiyan, Ojasvan, Sahamana Jayan, Abhijayan, Sudravina, Dravinodas, Ardrapavitra, Harikesa, Moda and Pramoda.[48] The three seasons mentioned are Agni, Surya and Candra.[49]

We have already said that the system of nomenclature beginning with Caitra was not current in the earlier Vedic period; its trace is not found in the Aitareya Brahmana, Taittiriya Samhita and Brahmana, and the Vajasaneyi Samhita and Brahmana. In order to arrive at the second system of nomenclature three stages were passed; first, the names of the Naksatras were known and the union of the moon with a certain Naksatra (lunar asterism) in which the moon was supposed to be full at that time was also known; secondly, the naming of the full moon after the asterism, e.g., Caitri Purnima, Vaisakhi Purnima etc., was next adopted; and thirdly, the month derived its name from the asterism in which the union took place, the moon being full. The names of the Naksatras or asterisms can, no doubt, be found in the earlier Vedic literature. The naming of the full moon after the asterism is also found in some places.[50] This much of the second system of nomenclature was reached even in the later part of the Vedic literature. The actual use of the names of the months in the order Caitra, Vaisakha, etc. was not in vogue at the time of the Vedic Samhita and Brahmana. The second system came into force long after the system beginning with Madhu was invented. In this system, we have already remarked, the months were named after the asterisms thus:

Month	*Lunar Asterism*	*Month*	*Lunar Asterism*
Magha	Magha	Sravana	Sravana
Phalguna	Uttaraphalguni	Bhadra	Purvabha-drapada
Caitra	Citra	Asvina	Asvini
Vaisakha	Visakha	Kartika	Krittika
Jyaistha	Jyestha	Margasirsa	Mrigasiras
Asadha	Purvasadha	Pausa	Pusya

It is difficult to state with accuracy the date when this adjustment of the months to the Naksatras took place. But mention of these names is frequently made in the *Manu Samhita,* in connection with the times when ceremonies or prescribed duties should be performed. A few extracts containing the names are given below:

"On the days of the conjunction and opposition let him (the father of a family) constantly make those oblations which are hallowed by the Gayatri, and those which avert misfortunes, but on the eighth and ninth lunar days of the three dark fortnights at the end of Agrahayana (Margasirsa) let him always do reverence to the Manes of ancestors.

"In the month of Asvina let him (the father of a family) cast away the food of sages, which he before had laid up, and his vesture, then become old, and his herbs and roots; the sun in the sign of Kanya (Virgin) must be shunned.

"Having duly performed the Upakrama (or ceremony for commencing the study of the Vedas) at the full moon of Sravana or Bhadra, let the Brahmin fully exert his intellectual powers, and read the Vedas during four months and one fortnight.

"Under the Lunar Asterism of Pusya, or on the first day of the bright half of Magha, and in the first part of the day, let him perform out of the town the ceremony Utsarga of the Vedas.

"Let him (the king) set out in his expedition in the fair month of Margasirsa, or about the month of Phalguna and Caitra, according to the number of his forces, that he may find autumnal or vernal crops in the land invaded by him."

From the above extracts it may be inferred that before the time when the *Manu Samhita* was composed or compiled, the names of the months after the asterisms were widely known, and that a long unmeasured anterior period must have elapsed before such a system, could have been so universally known and established.

It should be mentioned here that though the months were originally lunar,[51] their names being derived from the Naksatras in which the moon departing from a particular point was observed to be full, yet, later the months became solar, when the sun would be in the same Naksatra as before but diametrically opposite to the position of the moon. When did the solar months come into use? The Vedic literature contains the frequent mention of the synodic period or the lunar month, but nowhere in it is there any clear indication of the solar month. A solar month is the period the sun takes to pass through one division of the Zodiac which is divided into twelve parts. The twelve signs beginning with Mesa are not mentioned in the Vedas. Therefore, the solar months must have been in vogue after the Vedic period. But the period from one full moon to another or from one new moon to another which must obviously mean a lunar month is still in use in some parts of India. In fact, the solar month did not meet with so much favour as the solar year which has been universally accepted in India.

The next division of time was that of a month into two parts.[52] The *Atharvaveda* says, "Prajapati is the month, Krishna is his sun and Suklapaksa is his life." But the two parts of the month were called more often by the names, Purva-paksa and Apara-paksa[53]. In the Purva-paksa the Devas were created and in the Apara-paksa the Asuras. In the Purva-paksa the moon becomes newer and newer and in the Apara-paksa thinner and thinner. Such is the description of the two paksas in the Taittiriya Brahmana. Sukla is, therefore, the Purva-paksa when the moon becomes fuller and fuller, till it becomes the full moon, i.e., the period after a new moon to a full moon is called Sukla-paksa or the bright half; and the period after a full moon to anew moon is called Krishna or the dark half. The names Sukla and Krishna[54] replaced Purva and Apara long after the Vedic age, as they are not mentioned in the Vedic Samhita or Brahmana.

The days and nights of the two parts of a month had also different nomenclatures.[55] The days of the Purva or Sukla paksa were called samjnanam, vijnanam, prajnanam, janat, abhijanat, samkalpamanam, prakalpamanam, upakalpamanam, upaklptam, klptam sreyas, vasiyas, ayat, sambhutam and bhutam;[56] the nights were called darsa, drista, darsata, visvarupa, sudarsana, apyayamana, pyayamana, pyaya, sunrtera, apuryamana puryamana, purayanti, purna and paurnamasi.[57] The days of the Apara or Krishna paksa were called prastutam, vistutam, samstutam, kalyanam, visvarupam, sukram, amrtam, tejasvi, tejas, samrddham, arunam, bhanumat, maricimat, abhitapat and tapasyat[58]; the nights were called suta, sunvati, prasuta, suyamana, abhisuyamana, piti, prapa, sampa, tripti, tarpayanti, kanta, kamya, kamajata, ayusmati and kamadugha.[59] The days were neuter as "aha" is neuter and the nights were feminine as "ratri" is feminine. Here it may be noticed that the last night of the Purva or Sukla paksa was called Paurnamasi, but the last night of the Apara or Krishna paksa is not called Amavasya, it is called Kamadugha. It is not known why the latter name was chosen.

We have seen that a lunar month consists of thirty lunar days; but one lunar day is not equal to a savana day which period is the time intervening between one sunrise to the next. Thirty lunar days are, therefore, equivalent to twentynine and a half savana days. This period is divided into thirty equal parts and each part is called a tithi. A tithi is then shorter than a solar day which is called divasa (vara or vasara). Hence it is found that sometimes two tithis fall in one savana day; this occurs generally once in a month. What is a tithi? The Aitareya Brahmana says, "It is the period between the setting and the next rising of the moon."[60] The moon has fifteen phases or fifteen tithis, in fifteen the moon wanes and in the next fifteen the moon waxes.[61] The tithis are respectively called pratipad, dvitiya, tritiya, caturthi, pancami, sasthi, saptami, astami, navami, dasami, ekadasi, dvadasi, trayodasi, caturdasi, purnima or amavasya. Whether the tithi falls in the Sukla paksa or in the Krishna paksa is known by the addition of the word Sukla or-Krishna before the tithi, e.g., Krishna-pancami which means pancami of the Krishna paksa, Sukla-caturdasi which means caturdasi of the Sukla paksa, etc.[62] There is mention of another period called astaka in the Vedic literature.[63] The first eight days of a paksa is called astaka. There is paurnamasi astaka. and there is amavasya astaka[64] there are twelve paurnamasi astakas and twelve amavasya astakas[65]. The *Taittiriya Brahmana* says, "the first eight days after a Purnima, i.e, of a Krishna paksa comprise a period called astaka and the other days are called ahas; and the first eight days of a Sukla paksa comprise; a period called udrsta and the other days are called ahas."[66] In this connection it will be interesting to cite here the explanation given by the Brahmanas regarding the denomination of amavasya. The Satapatha Brahmana says, "The gods said, "Nothing but Soma will satiate him; let us prepare Soma for him!" They prepared Soma for him. Now this king Soma, the food of the gods, is no other than the moon. When he (the moon) is not seen that night either in the east or in the west, then he visits this world and here he enters into the waters and plants. He is indeed a treasure for the gods, he is their food. And since during that night he here dwells together (ama vas) therefore that (night of new moon) is called amavasya"[67] The *Aitareya Brahmana* says, "In an amavasya the moon enters the sun. In fact, the moon is born of the sun.[68] There is mention of two special nomenclatures for the amavasya and the purnima to which we must refer. In the *Matsya Purana* and in the Vayu Purana it is said that in an amavasya the moon and the sun coming together looks on one another; hence amavasya is called darsa. The purnima is called parvan. The *Rigveda* also mentions another set of names for amavasya and purnima viz., Raka and Sinivali.[69] There are two other names mentioned in the Brahmanas, viz., anumaṭi and kuhu. Paurnamasi if fallen on the tithi caturdasi is called anumati, that falling on the tithi pratipad is called raka; amavasya, falling on the tithi caturdasi is called sinivali and that falling on pratipad is called kuhu.[70] It is not possible to ascertain how some of these names were coined and what special significance they bore at the time of their first use. In vain should we now search for their previous implication and whatever meaning some of these names had got before, they surely lost their former import long ago.

The next division of the calendar that is of great importance is the week consisting of seven days. In the Vedic literature we get no separate names for the seven days; there is only one common name for the seven, and that is called vasara.[71] Sayana has explained vasara as divasa and has given the literal meaning as "nivasaka" or "nivasasya hetubhuta." It means the time when the sun lives. There is another passage in the *Rigveda* where the word vasara has been used in the literal sense, namely jagadvasaka.[72] In the Atharva Jyotisa the seven deities presiding over the seven days of a week are mentioned; they are Aditya, Soma, Bhauma, Budha, Brihaspati,

Bhargava and Sanaiscara.[73] Invariably a connection is sought to be established between the days and the planets. In the *Yajnavalkya Smriti* nine planets are mentioned in connection with the sacrifice of the planets: they are Surya, Soma, Mahiputra, Somaputra, Brihaspati, Sanaiscara, Rahu and Ketu.[74] But nowhere the names of the seven days over which the seven planets preside are given. Even in the *Mahabharata* the seven days are not given by names. There is only a solitary mention of the word "vara" in connection with the anecdote of the Raksasa in the Ekacakranagara.[75] There "vara" has been used in the sense of "a single time," "once" and "not twice." The names of the seven days came into vogue long after the Vedic or the Pauranic age. Though the planets which were supposed to preside over the seven days were mentioned by names in several places, yet the naming of the seven days after the respective presiding planets was of a much later period. It is believed that the Hindus got this practice of naming the seven days of the week, Ravi (or Surya), Soma, Mangala, Buddha, Brihaspati, Sukra and Sanaiscara, after the presiding planets from the Chaldeans who had from very early times this usage current amongst them.[76]

The next division of time is the divasa or day. We have seen that the Saura or solar months, each consisting of 30 Saura days or 30 degrees, are of unequal length reckoned in mean solar days on account of the unequal motion of the sun in the ecliptic, but the aggregate is equal to the sidereal year, which in the *Surya Siddhanta* is reckoned to be 365 days 6 hours 12 minutes 36.56 seconds and the mean Saura, or solar month, would, therefore, be 30 days 10 hours 31 minutes 35 seconds.[77] The Saura month of the greatest length is Asadha, consisting of 31 days 14 hours 39 minutes 7 seconds; and the least is Pausa which is 29 days 8 hours 21 minutes 7 seconds. We have already said that the lunar month and the lunar day were at first used which subsequently gave place to the solar month and the solar day. Now what is a Savana or solar day?

The *Surya Siddhanta* says, "The time from one rising of the sun to the next is called Savana or a terrestrial day, from this the number of terrestrial days in a kalpa is determined; by these days the time of a sacrifice is calculated." The day is again divided in various ways. In some places it is divided into two parts, viz., purvahna (forenoon) and aparahna (afternoon); in some places in three parts, forenoon, mid-day (madhyahna) and afternoon.[78] The Satapatha Brahmana says, "The forenoon, doubtless, belongs to the gods; the mid-day to men, and the afternoon to the fathers."[79] The day is some times divided into four parts, forenoon, mid-day, afternoon and evening (sayam); it is again divided into five parts,—dawn, samgaba (the time when the cows are driven together for milking), mid-day, afternoon and evening (sayam).[80] The last kind of division of a day is into muhurtas, a day is divided into fifteen muhurtas and so also a night.[81] The day and night consists of thirty muhurtas, a muhurta = about 48 minutes or 4/5 of an hour. "One year = 360×30= 10,800 muhurtas which is just the amount of eighties of which the three Vedas were to consist; for *Rigveda* consists of 36×12,000 syllables, Yajus 36×8000 and Saman 36×4000 syllables, the total is 2×432,000 =80+10,800."[82] The fifteen divisions of a day into muhurtas have not fifteen different names, so also the fifteen divisions of a night into muhurtas; again the divisions have got different names for days and nights of the Sukla and Krishna paksas. The fifteen divisions of a day of the Sukla paksa were called citra, ketu, prabhan, abhan, sambhan, jyotisman, tejasvan, atapan, tapan, abhitapan, rocana, rocamana, sobhana, sobhamana, and kalyana;[83] the fifteen divisions of a night of the Sukla paksa were called data, pradata, ananda, moda, pramoda, avesayan, nivesayan, samnesana, samsanta, santa, abhavan, prabhavan, sambhavan, sambhuta and bhuta.[84] The fifteen divisions of a day of the Krishna paksa were called savita, prasavita, dipta, dipayan, dipyamana, jvalan, jvalita, tapan, vitapan, santapan, rocana, rocamana, sumbhu, sumbhamana and vama;[85] the fifteen divisions of a night of the Krishna paksa were called abhisasta, anumanta, ananda, moda, pramoda, asadayan, nisadayan, samsadana, samsanna, sanna, abhu, vibhu, prabhu, sambhu and bhuva.[86]

Again each muhurta was subdivided into fifteen pratimuhurtas.[87] These fifteen pratimuhurtas had also different names; they were idanim, tadanim, etarhi, ksipram, ajiram, asu, nimesa, phanodravan, atidravan, tvaran, tvaramana, asiyan and Java.[88] There were further divisions into kala, kastha and Nimesa.[89] In the *Mahabharata* the division of a day and a night into kastha, kala, muhurta and lava is mentioned.[90] There were also other divisions, called nadika (2 nadikas = 1 muhurta), pala, mashka, drona, adhaka etc.[91] The Atharva Jyotisa gives the following table:—

12 nimesa = 1 lava	30 truti (or 15 pratimuhurtas)= 1 muhurta
30 lavas = 1 kala	30 muhurtas = 1 day and night.
30 kala = 1 truti	

The above statement shows how much careful the Hindus were in dividing the periods into the minutest particles of time. Though most of them have now lost their former importance and are almost useless in practical work, yet it testifies to their utmost endeavour to be precise and correct to the nearest approximation.

Now we proceed to discuss an important question in connection with the calculation of time. When was the day reckoned to have its beginning? The Vedic and the Pauranic literature maintained that the day began with sun-rise.[92] But different theories on this question were advanced by the later Hindu astronomers. Varahamihira says, "Aryabhata maintains that the beginning of the day is to be reckoned from mid-night at Lanka, and the same teacher again says that the day begins from sunrise at Lanka."[93] But in the present text of the Aryabhatiya we find only the latter theory, namely, that the day begins from sunrise.[94] The personal opinion of Varahamihira is that the day begins from mid-day at Lanka. Brahmagupta holds the same view and says that mid-day at Lanka corresponds to sunset in the city of the Yavanas. Latacarya maintains that the day begins from sun-set in the city of the Yavanas, and, therefore, he belongs to the same school as Varahamihira and Brahmagupta. Bhattotpala, the commentator of Brahmagupta, speaks of four kinds of day-beginnings, namely, from sun-rise, from mid-night, from mid-day and from sun-set. Simhacarya maintains that the day begins from sun rise at Lanka; but his preceptor says that the day begins ten muhurtas after the fall of night, i.e.. 10×12/15 or 8 hours after night-fall.[95] Varahamihira discusses at a considerable length all these theories in his Pancasiddhantika, chap. xv.

The division of the Calendar into a year, months, paksas and days evolved gradually but the system was almost complete in the Vedic and the Pauranic ages. Slight modifications were introduced later on by the Hindu astronomers and this modified calender has been since then in vogue in India. Only a few slokas have been devoted by Bhaskara in his *Siddhanta Siromani* for its explanation, some extracts of which are given below:[96]

"If a star and the Sun rise simultaneously (on any day), the star will rise again (on the following morning) in 60 sidereal ghatikas; the sun, however, will rise later by the number of asus (sixths of a sidereal minute), found by dividing the product of the sun's daily motion (in minutes) and the asus which the sign, in which the sun is, takes in rising, by 1800 (the number of minutes which each sign of the ecliptic contains in itself). This time thus found added to the 60 sidereal ghatikas forms a true terrestrial day or natural day. The length of this day is variable, as it depends on the sun's daily motion and on the time (which different signs of the ecliptic take) in rising, (in different latitudes, both of which are variable elements).[97] A sidereal day consists invariably of 60 sidereal ghatikas; a mean Savana day of the sun or terrestrial day consists of that time with an addition of the number of asus equal to the number of the sun's daily mean motion (in minutes).[98] Thus the number of terrestrial days in a year is less by one than the number of revolutions made by the fixed stars.

"The length of the solar year is 365 days, 15 ghatikas, 30 palas, 22½ vipalas reckoned in Bhumi Savana or terrestrial days. The 1/12 th. of this is called a Saura or solar month, viz, 30 days, 26 ghatikas, 17 palas, 31 vipalas, 52½ pravipalas. Thirty Savana or terrestrial days make a Savana month.[99] The time in which the moon after being in conjunction with the sun, completing a revolution with the difference between the daily motion and that of the sun again overtakes the sun, which moves at a slower rate, is called a lunar month. It is 29 days, 31 ghatikas, 50 palas in length.

"An adhimasa or additive month which is lunar, occurs in the duration of 32½ Saura or solar months found by dividing the lunar month by the difference between this and the Saura month. As a mean lunar month is shorter in length than a mean Saura month, the lunar months are, therefore, more in number than the Saura in a Kalpa. An avama or subtractive day which is Savana occurs in $62\frac{1}{11}$ tithis (lunar days) found by dividing 30 by the difference between the lunar and Savana month."[100]

We shall close our discussions by a reference to the Jaina calendar which differs in some respects from the Vedic and Pauranic calendar. The only work on Jaina astronomy now available is the Suryaprajnapti written by Mahavira. It is written in Prakrit, but its commentary by Malayagiri is written in classical Sanskrit. The probable date of the Suryaprajnapti synchronises with that of the Jyotisa Vedanga. The following will clearly explain the formation of the calendar in the Suryaprajnapti:—

The moon moves and unites 67 times with Abhijit in a Yuga of 5 years. The sun comes in contact five times with the same star in a Yuga.

The names of the months are: —

1.	Sravana	Abhinanda
2.	Bhadrapada	Supratistha
3.	Asvayuja	Vijaya
4.	Kartika	Pritivardhana
5.	Margasirsa	Sreyan
6.	Pausya	Siva
7.	Magha	Sisira
8.	Phalguna	Haimavan
9.	Caitra	Vasanta
10.	Vaisakha	Kusumasambhava
11.	Jyaistha	Nidagha
12.	Asadha	Vanavirodhi

Years:—(1) Naksatra-samvatsara = 12 Naksatra-masas=

$$12 \times 27\frac{21}{67} \text{ days} = 327 \text{ days} + \frac{51}{67} \text{ days}$$

(2) Yuga-samvatsara (cyclic year) = 5 years

(3) Pramana-samvatsara. (4) Saturn-year.

The first is of 12 kinds, as Sravana, Bhadrapada, etc.; when Jupiter completes the whole circle of constellations once, it is called, Naksatra-samvatsara of 12 years.

$$\text{Lunar year} = 29\frac{32}{62} \times 12 = 354 \text{ days} + \frac{12}{62} \text{ days}$$

Intercalary Lunar year = $383 + \frac{44}{62}$ days

Saura or solar year= 12 × 30½ = 366 days

Thus, once in 30 solar months there will be one intercalary lunar month. Hence in a Yuga of 60 solar months there will be two intercalary lunar months. Each lunar month contains two parvas. Therefore, a lunar year contains 24 parvas, and an intercalary year 26 parvas.

The Pramana-samvatsara is of five kinds: Naksatra (sidereal), Rtu (seasonal), Candra (lunar), Aditya (solar) and intercalary lunar. The sidereal and lunar years have been just explained. The Rtu and Aditya-samvatsaras will now be explained:

2 Ghatikas=1 Muhurta	15 Days = 1 Paksa
30 Muhurtas=1 Day and Night	2 Paksas = 1 Month
	12 Months = 1 Year

The year of 360 days and nights is a Rtu-samvatsara. This has two more names, Karma-samvatsara and Savana-samvatsara; Karma = work (laukikavyavahara). Hence that year which is prominently observed by workmen is so called. Karma month has no fraction and facilitates work and worldly transaction; the rest have fractions and so in usage it is difficult to understand. Savana means engagement in work. Hence that year which is chiefly agreeable to work is Savana year. The year of 360 days is called Karma and also Savana year. Similarly, the solar year is the time taken by the rainy and other seasons for completion of one revolution. It is, however, usual to assign 60 days to each of the seasons. Still each one of them has 61 days. Hence the solar year contains 366 days. In a yuga there are three ordinary lunar years of $354 \frac{12}{62}$ days and two intercalary years of $383 \frac{40}{62}$ days. Hence in a yuga there are 62 lunar months, 67 Naksatra months. The lunar year and also the solar year commence at the same point or day and close at the same point or day once in every cycle of 30 years which is equal to 6 cycles of 5 years each. For the lunar year gains 6×2 months and thus completes one whole intercalary year. Similarly, the solar, the Savana or seasonal, the lunar, and the Naksatra years begin on the same day and close on the same day or simultaneously begin and close once in 12 cycles of 5 years each, i.e., 60 years. In a similar way the intercalary lunar year, the solar, the Rtu or Savana, the lunar and the Naksatra years will simultaneously begin and close once in a grerat cycle of 156 cycles of 5 years each; for 156×5 years are equal to 744 intercalary years, 780 solar, 793 Rtu, 806 lunar and 871 Naksatra years.

Next the measure of solar and other month is given: (a) A solar year is equal to 366 days, hence one solar month is $\frac{366}{12} = 30\frac{1}{2}$ days; (b) a Karma-samvatsara = 360 days; hence one Karma-month = $\frac{360}{12} = 30$ days; (c) a lunar year $= 354\frac{12}{62}$ days, hence one lunar month $\frac{354\frac{11}{62}}{12} - 29\frac{32}{62}$ days; (d) a Naksatra year = $327\frac{51}{67}$ days, hence one Naksatra-month $\frac{327\frac{51}{67}}{12} = 27\frac{21}{67} =$ days; (e) an intercalary lunar year = $383\frac{44}{61}$ ·days, hence one intercalary month = $\frac{383\frac{44}{61}}{12} = 31\frac{121}{124} =$ days.

In a yuga or cycle of 5 years or 1830 days, there are 60 solar months, or 61 Savana months, or 62 lunar months or 67 Naksatra months or 57 intercalary months, 7 days, $11\frac{23}{62}$ muhurtas.[101]

Again one lunar month is divided into two parts, the white half contains $442\frac{46}{62}$ muhurtas and the dark half also

$442\frac{46}{62}$ muhurtas. A tithi or lunar day is equal to $\frac{61}{62}$ parts of a day. Hence a day being divided into 30 muhurtas, a tithi will be equal to $\frac{61}{62} \times 30$ muhurtas = $29\frac{32}{62}$ muhurtas. The tithi are of two kinds: (1) day tithis and night tithis; both kinds are divided into a week of five lunar days, called (a) Nanda, (b) Bhadra, (c) Jaya, (d) Tuccha, (e) Purna, in the case of day tithis; and (a) Ugravatl, (b) Bhogavati, (c) Yasomati, (d) Sarvasiddha and (e) Subhanamni, in the case of night tithis. Thus three weeks of day tithis and three weeks of night tithis will make fifteen complete lunar days. The above is a short summary of the division of time mentioned in Mahavira's Suryaprajnapti.

This finishes a general discussion about the formation of the Hindu Calendar. The Hindus from very early times tried to get a detailed calculation of the year, month, paksas, day and the sub-divisions of a day. The calendar was in most respects perfected in the Vedic and the Pauranic age, though some modifications were introduced by the later Hindu astronomers.

Reference

1. Whitney's notes to the *Surya Siddhanta*, 1, 13.
2. Tilak's *Orion*, p. 11.
3. संवत्सरः प्रजापति । प्रजापतिर्यज्ञः ।—*Ait. Brahmana*, ii, 17: संवत्सरी वै प्रजापतिः प्रजापतिर्यज्ञः ।— *Satapatha Brahmana*, ii, 2, 2, 4. Also Ait. Br., iv, 22; *Satapatha Br.*, xi, 1, 1, 1; 2, 71.
4. Brennand's *Hindu Astronomy*, p. 58.
5. The word मास is derived from चन्द्रमस्, and month at first meant lunar month only. The English words "moon" and "month" are similarly derived. The word "मास" also means चन्द्र as in सूर्यामासा (*Rigveda*, 8, 94, 2). Another synonym for the moon is मासकृत्.

6. *Tait. Brahmana,* 1,5, 2, 1.
7. *Satapatha Brahmana,* 4,6,1,11; 6,2,1,26; 6,2,2,5 and 12. Also vide Prof. Jogesh Chandra Roy's book, *Our Astronomy and Astronomers,* p. 10.
8. Surya Siddhanta, 1, 13-21.
9. The gods are supposed to reside on Mount Meru under the North Pole, where the day lasts for six months. The Asuras are said to reside in the South Pole.
10. The Sandhya and Sandhyamsa are the dawn and evening twilight, and as the days of mortals have these, so from analogy those of the gods had been likewise.
11. It will be interesting to note here that mn + r = 10,000, is just one of the important indeterminate equations of the Hindus for the solution of which their early writers sometimes challenge each other in grotesque language. Vide the seventh and eighth sections of chapter XVIII, *Brahmasphuta Siddhanta.*
12. Brennand's *Hindu Astronomy,* p. 182.
13. *Surya Siddhanta,* 1, 37.
14. *Surya Siddhanta,* 1, 57 and 58.
15. Brennand's *Hindu Astronomy,* p. 36.
16. *Rigveda,* x, 72, 2; i, 103, 4; v, 52, 4; v, 73, 3; 1, 158, 6; vi, 8, 5.
17. *Ibid.,* x, 97, 1; *Tait. Samhita and Vajasaneyi Samhita,* 12,75; Sayana names them as Krta, Treta and Dvapara.
18. *Tait. Samhita,* iv, 3,3.
19. *Vajasaneyi Samhita,* 30,18.
20. *Tait. Brahmana,* iii, 4, 1; *Ait. Brahmana,* 33, 15.
21. *Rigveda,* vii, 103, 7,8.
22. *Vajasaneyi Samhita,* 26, 45; 30, 15 and *Tait. Br.,* iii, 4, i.
23. *Tait. Br.,* 1, 4, 10.
24. *Tait. Br.,* iii, 10, 4.
25. S. B. Dikshit's *Bharatiya Jyotih Sastra,* p. 37.
26. *Tait. Br.,* i, 4, 10.
27. Vide the article on "The Indian cycle of 60 years" by Davis, *Asiatic Researches,* vol. III, Calcutta, where the order in which

these years are arranged is given in a very illuminating figure.

28. Brennand, *Hindu Astronomy,* p. 22.
29. Dr. Schrader, in his Prehistoric Antiquities of the Aryan Peoples, part IV, chap. VI, p. 305, also makes a similar observation. He holds, on philological grounds, that the conception was already formed in the primeval period by combining into one whole the conception of winter and summer which he believes to be the two primeval seasons. Vide Orion, p. 13.
30. *Rigveda,* i, 164, 48.
31 Savana is derived from √su to extract (Soma-juice) in a sacrifice, and means literally a sacrificial day.
32. *Ait. Br.,* ii, 17; *Tait. Samhita,* ii, 5,8,3; *Rigveda,* i, 164,48; *Satapatha Br.,* ii, 2, 2, 35; vii, 2, 4, 9, 16, 22; ii, 2, 2, 4.
33. *Tait. Samhita,* vii, 5, 7, 1; *Tandya Br.,* v, 10.
34. *Tait. Samhita,* i, 4, 14; *Vaj. Sam.,* 7, 30; *Satapatha Br.,* ii, 2, 3, 27; iii, 6, 4, 24; 4, 5, 23; vi, 2, 2, 29; vii, 7, 1, 28. As regards the twelve hallowed (intercalary) night, vide *Rigveda,* iv, 33, 7; *Atharvaveda,* iv, 11, 11, *Tait. Br.,* i, 9, 10. For other allusions to intercalary months in the Vedic texts, see Weber, Naksatra, ii, p. 336.
35. *Rigveda,* i, 25, 8.
36. Brennand, *Hindu Astronomy,* p. 60.
37. Vide *Satapatha Br.,* ii, 2, 2, 5.
38. Vide *Satapatha Br.,* 11, 2, 2, 35.
39. *Tait. Samhita,* v, 6, 7; vii, 5, 1.
40. Vide Burgess and Whitney, *Surya Siddhanta,* i, 13, n. "It is, however, not the tropical solar year which we employ, but the sidereal, no account being made of the precession of the equinoxes."
41. *Tait. Samhita,* i, 6, 7; vii, 5, 6. Vide also S.B. Dikshit's *Bharatiya Jyotih Sastra,* p 41, 42.
42. *Satapatha Br.,* iv, 3, I, 14-19.

43. *Tait. Sam.*, 1, 4, 14; iv, 4, 11. *Vajasaneyi Sam.*, Viii, 30; xiii, 25; xiv, 6, 15; xvi, 27; xxii, 31.
44. *Vajasaneyi Sam.*, vii, 30; xxii, 30.
45. *Ait. Br.*, 3, 1.
46. *Tait. Br.*, iii, 10, 9.
47. *Tait. Br.*, iii, 10, 1.
48. *Tait. Br.*, iii, 10, 1.
49. *Ibid.*
50. *Tait. Br.*, vii, 4, 8; 1,2, 8. *Satapatha Br.*,—Esa ha samvatsarasya prathama ratrir ya phalguni purnamasi.

    Tandya Br., v, 9; *Gopatha Br.*, vi, 19; *Samkhayana Br.*—Ya vaisa phalguni purnamasi samvatsarasya prathama ratrih. *Samavidhana Br.*, ii, 4.
51. It will be interesting to mention here that the lunar month has a different beginning in different parts of India. In Bengal it begins at the full moon or Purnima—the midnight of the Pitrs.
52. Some say that this division was done by way of analogy as the day of the Pitris was divided into two parts, by mid-day and mid-night, when the moon was in opposition or conjunction, so the month was divided into two parts, the bright half and the dark half, or as they are called, the Suklapaksa and Krishnapaksa, each part or paksha consisting of 15 lunar days or tithis.
53. *Tait. Br.*, ii, 2, 3, 1; iii, 10, 4, 1.
54. Sukla and Krishna are mentioned in the Atharva Jyotisa.
55. *Tait. Br.*, iii, 10, 10,2
56. *Ibid.*, iii, 10, 1, 1.
57. *Ibid.*, iii, 10, 1, 1.
58. *Ibid.*, iii, 10, 1, 2.
59. *Ibid.*, iii, 10, 1, 2 and 3.
60. "Yam paryastani iyad abhyudiyad iti sa tithih."—*Ait. Br.*, 32, 10.
61. *Tait. Br.*, i, 5, 10.
62. *Samavidhana Brahmana*, ii, 6, 8; iii, 3.

63. *Tait. Br.*, i, 5, 12.
64. *Tait. Br.*, iii. 11, 1, 19.
65. *Tandya Br.*, x, 3. 11.
66. *Tait. Br.*, i, 8, 10, 2; *Tandya Br.*, xviii, 11. 8.
67. *Satapatha Br.*, i, 6, 4, .5; vi, 2, 2, 16; xi, 1, 5, 1.
68. *Ait. Br.*, 40, 5.
69. *Rigveda*, ii, 32.
70. *Ait. Br.*, 32, 10; *Gopatha Br.*, vi, 10.
71. *Rigveda*, viii, 6, 30.
72. *Rigveda*, viii, 48, 7.
73. *Atharva Jyotisa*, Sl. 93. The Atharva Jyotisa consists of 162 slokas and 14 chapters.
74. *Acaradhyaya*, verse 295.
75. *Mahabharata*, Adi Parva, chap. 160.
76. Vide S. B. Dikshit's *Bharatiya Jyotihsastra*, pp. 138 f.
77. *Surya Siddhanta*, xiv, 12-14.
78. *Tait. Br.*, iii, 12, 9, 1.
79. *Rigveda*, ii, 4, 2, 8.
80. *Rigveda* v, 76, 3; Tait. Br., i, 5, 3; *Satapatha Br.*, ii, 2, 3, 9; *Atharvaveda*, ix, 6, 46.
81. *Tait. Br.*, iii, 10, 9; *Satapatha Br.*, x, 4, 2, 8.
82. *Satapatha Br.* x. 4, 2. 9.
83. *Tait. Br.*, iii. 10. 1.
84. *Tait. Br.*, iii. 10. 1. 1. 2.
85. *Tait. Br.*, iii. 10. 1. 1. 2.
86. *Tait. Br.*, iii. 10. 1. 3.
87. *Tait. Br.*, iii. 10. 1. 4.
88. *Tait. Br.*, iii. 10, 9. 9.
89. *Narayana Upanisad*, Anukramanika, verse 1.
90. *Mahabharata*, Santiparva, Chap. 7, verse 21; Chap. 36 verse 14.
91. *Yajurveda Jyotisa*, verses 16, 17; *Brihat Samhita*, Chap. 23, verse 2, Vide also a table given in Lilavati.

92. There might have been a section at the time of the later Brahmanas, who maintained that the day began with midday.
93. *Pancasiddhantika,* (Thibaut and Dvivedi's edition) chap. xv, verse 20.
94. *Aryabhatiya,* Gitikapada, chap. x, verse 2.
95. *Pancasiddhantika,* chap, xv, verse 20.
96. *Siddhanta Siromani,* Goladhyaya, chap. IV, v, 5-14.
97. "Had the sun been moving with uniform motion on the equinoctial, each minutes of which rises in each asu, the number of asus equal to the number of the minutes of the sun's daily motion, being added to the 60 sidereal ghatikas, would have invariably made the exact length of the true terrestrial day as Lalla and others say. But this is not the case, because the sun moves with unequal motion on account of its being oblique to the equinoctial. Therefore, to find the exact length of the true terrestrial day, it is necessary to determine the time which the minutes of the sun's daily motion take in rising and then add this time to 60 sidereal ghatikas. For this reason, the terrestrial day determined by Lalla in his Sisyadhivrddhida and also by some others is not a true but a mean day."—Footnotes by Bapudev Sastri on Goladhyaya chap. IV, verses 5, 6).
98. Compare *Surya Siddhanta,* chap. XIV, verse 15, which says, "The only invariable astronomical unit is the sidereal day, or the time of one apparent revolution of the sphere of the stars about the earth."
99. Here a solar year consists of 365 days, 15 ghatikas, 30 palas 22½ vipalas, i.e., 365 days. 6h. I2m. 9s. But in the Surya Siddhanta, the length of the year is 365 days, 15 ghatikas, 31 palas, 31.4 vipalas, i.e., 365 days, 6 h. 12m. 36.56 s.
100. For further explanation of these verses see the footnotes by Bapudev Sastri on Siddhantasiromani, Goladhhyaya, chap, iv, verses 10-12.
101. For an intercalary month = $31\frac{121}{124}$ days. Hence $\frac{1830}{31\frac{121}{124}} = \frac{226950}{8565}$ = 57 months, 7 days and $11\frac{23}{62}$ muhurtas.

[Courtesy: Sukumar Ranjan Das].

3

The Mayan and Hindu Calendars: Comparison and Correlation

If there was any relationship between the calendars of the Mayas of Mexico and that of our Hindu calendar?

The following quotation from the article 'Mexico' in the *Encyclopaedia Britannica* (13th ed.) on the prehistoric civilization of the Mayas leads one to the belief that there must be some connection between the Mayan and Hindu calendars. '...So that the general tendency among anthropologists has been to admit a common origin, however remote between the tribes of Tartary and of America. ... Humboldt also discussed the Mexican doctrine of four ages of the world belonging to water, earth, air and fire and ending respectively by deluge, earthquake, tempest and conflagration. The resemblance of this to some versions of the Hindu doctrine of the four ages or Yugas is hardly to be accounted for except on the hypothesis that the Mexican theology contains ideas learnt from Asiatics. Among Asiatic points of resemblance to which attention has been called is the Mexican belief in the nine stages of heaven and hell, which nothing in nature could suggest directly to a barbaric people but which corresponds to the ideas of successive heavens and hells among Brahmins and Buddhists who apparently learnt it (in common with our ancestors) from the Babylonian Greek theory of successive stages or concentric planetary spheres belonging to the planets, etc. The Spanish

Chronicles also give account of a Mexican game called 'potolli' played at the time of the conquest with coloured stones moved on the squares of a crosslegged figure according to the throws of beans marked on one side; the description of this rather complicated game corresponds closely with the Hindu backgammon called Pachisi ..... It was believed that the destruction of the world, which after the Hindu manner the Mexicans held to have already taken place three or four times, would happen again at the end of a cycle. As the time (the end of the 52 year cycle) drew near the anxious population cleared their houses and put out all fire and on the last day after sunset the priests, dressed in the garb of gods, set out in procession for the hill of Huixachtle there to watch for the approach of the Pleiades to the zenith which gave the auspicious signal for the lightning of the new fire. .... Marriage depended much as they do still in the East, on the comparison of the horoscopes of the pair to ascertain if their birth signs were compatible. Old women were employed as go-betweens, and the marriage ceremony was conducted by a priest who after moral exhortations united the young couple by tying their garments together by a knot, after which they walked seven times round the fire, casting incense into it; after the performance of the marriage ceremony, the pair entered together on a four days' fast and penance before the marriage was completed.' The very early and later funeral rites also closely resemble those of the Hindus. It is to be noted that the lightning of the new fire was begun on the approach of the Pleiades (the Hindu Nakshatras *Krittika)* on the zenith (or, on the horizon?). The statement in the *Satapatha Brahmana* '..कृत्तिकासु अद्यौ आदधीत। एताः ह वै प्राच्यै दिशः न च्यवन्ते।...' 'One should light the fire on the appearance of the Krittikas. These never swerve from due East . . .' The custom of the Mayas and the importance of the Krittika (the Sapta Matrika, the seven mothers, wives of the Saptarshis—the Great Bear)—the Mataliki or Matariki of the Maoris of New Zealand deserve special attention. There are numerous evidences in support of the fact that these Mayas and the Maoris of New Zealand migrated from India in prehistoric times.

We now come to the astronomical verification that the epoch of the Mayan and the Hindu calendars are identically the same. J.E. Thompson in his 'A correlation of the Mayan and European Calendars ' (publication No. 241 of the Field Museum of Natural History, Chicago) calculated the epoch of the Mayan calendar and fixed this to August 13 (Gregorian), 3113 B.C. But it should be noted that his conversion of the Mayan dates into their Julian equivalents by the addition of the Ahau equation of 584285 days does not lead to 3113 B.C., but the year—3113 of the astronomers or the year 3114 B.C. Now this year 3114 B.C. is twelve years prior to the epoch of the Hindu Kali Yuga era, which is also the epoch of the sixty year cycle of Jupiter by the Telinga reckoning of Southern India. But the examination of several Mayan dates with this epoch of 3114 B.C. does not lead to any satisfactory result Mr. Thompson submitted six dates from the Dresden Codex after converting them into their Julian equivalents by the addition of his supposed Ahau equation of 584285 to the U.S. Naval Observatory. The importance of the dates was supposed to be due to some phenomena connected with the planets Mars, Jupiter and Saturn. But the information received from Captain W.S. Eichelberger, the Director of the American Nautical Almanac who calculated with the help of tables of Dr. Paul V. Neugebauer shows clearly that none of the dates has any connection with any phenomenon connected with the supposed planets. The epoch of the Mayan dates may have some connection with the Hindu calendar—the epoch of the Kali Yuga era, 18th February, 3102 B.C. (the day begins from midnight, so that 18.00 February, Ujjaini time was midnight of February 17, Julian day number 588465). Julian day number 588466. The dates in the Dresden Codex in the long count were converted to their Julian equivalents by applying the Kali Yuga equation of 588466 with the following results:—

	Ling count dates.		Julian equivalents.
(1)	8-16-14-9-3	= 1860889	= Nov. 2, A.D. 382.
(2)	8-16-14-11-5	= 1860931	= Dec., 14, A.D. 382.
(3)	9-13-10-15-14	= 1981980	= May 14, A.D. 714.
(4)	9-18-0-12-9	= 2014315	= Nov. 23, A.D. 802.
(5)	9-19-7-2-14	= 2023840	= Dec. 21, A.D. 828.
(6)	9-19-7-17-7	= 2024133	= Oct. 10, A.D. 29.

On the first date, November 2, A.D. 382 the planets Venus and Saturn were in geocentric opposition and the planet Mercury attained the greatest western elongation.

On the second date, December 14, A.D. 382 the planets Mars and Jupiter were in geocentric opposition. Mercury—superior conjunction; and Jupiter rose heliacally on this date. [All the heliacal phenomena have been computed for lat. I7°N.]

On the third date, May 14, A.D. 714 the following planetary phenomena occurred: Jupiter in opposition with the sun. Only two days before retrogradation ended and direct motion of Saturn began and Venus attained the greatest brilliancy.

On the fourth date, November 23, A.D. 802, Jupiter and Venus were in mean heliocentric opposition.

On the fifth date, December 21, A.D. 828 the following phenomena occurred: Jupiter and Saturn in conjunction; Venus and Jupiter in geocentric opposition; and Mercury set heliacally in the east.

The sixth date as given is 9-19-7-17-7 equivalent to October 10, A.D. 829. But on calculation no planetary phenomenon is found to have occurred on this date. It seems to me that the date may either be 9-19-7-11-7 or 9-19-7-16-7. The former date is equivalent to June 12, A.D. 829 on which date Jupiter was in geocentric conjunction with Saturn. The latter date is equivalent to September 20, A.D. 829 on which date Venus was in geocentric conjunction with Saturn and the autumnal equinox occurred the previous day.

There is another date in the Dresden Codex about which Mr. Thompson remarks 'The Dresden Codex seems to indicate that a lunar count began on the day 9-16-4-10-8 or a day earlier or later. Presumably the date either of a new moon or less likely a full moon and possibly an eclipse date. Dr. Teeple's elucidation of glyphs C. D and E of the lunar series has established the fact that this date was actually the basis from which the lunar count was reckoned.' Now referring this date 9-16-4-10-8 to the epoch of the Kali Yuga era we come to April 20, A.D. 767. On the 18th of April A.D. 767 there was a total lunar eclipse the passing phase of which was visible from Mexico for a short time in the evening. A fortnight before on April 3, there was a total solar eclipse and a fortnight later on May 2, there was the chance of solar eclipse.

I now come to a few dates found in J. E. Thompson's *'The Civilization of the Mayas'* in the chapter entitled 'the Maya Collection of Field Museum ' (pp. 94-102).

The date in Altar Q from Copan has been read as 6 Caban 10 Mol corresponding to the long count date 9-16-12-5-17. This date is equivalent to December 8, A.D. 774. This was a new moon day on which Mercury was stationary in the morning and two days later Jupiter set heliacally. This is a date most frequently recorded at Copan but Mr. Thompson could not find out the exact significance of this date.

Six other dates have been read in this altar Q at Copan. The first one is 5 Caban, 15 Yaxkin corresponding to the long count date 9-15-6-16-7 equivalent to November 29, A.D. 749. Full moon occurred the previous day. Jupiter and Saturn approached within a degree on this date and the sun entered the ancient Hindu constellation Sagittarius.

The second date is 8 Ahau, 18 Yaxkin, only 3 days later than the previous date, i.e., December 2, A.D. 749. The importance of this date lies in the fact that on this date Mercury was in conjunction with Jupiter and Saturn and Saturn rose heliacally on this date.

The third date is 5 Ben, 11 Muan, i.e., 153 days later than the previous date, equivalent to May 4, A.D. 750. On this day the planets Venus and Mercury were in conjunction.

The fourth date is 6 Ahau, 13 Kayab written in the long count 9-17-5-0-0. This date is equivalent to June 6, A.D. 787. This was a full moon day on which planet Mars set heliacally. Only 2 days later Mercury was in superior conjunction with the sun. In his '*Civilization of the Mayas*', (p. 95) Mr. Thompson remarks, ' In F 3 there is the glyph for the observation of the sun at the horizon. Possibly this is to call attention to the fact that the important hotun date is only a few days off the winter solstice.' Curiously the fourth date (the important hotun date) June 6, A.D. 787 is only 12 days previous to the Summer solstice. Hence the indication in the glyph for the observation of the sun relates to a coming phenomenon, with Thompson's epoch winter solstice occurred 7 days before the date deduced by him (25th December, Jul. A.D. 775).

The last date is 3 uinals and 4 kins to be added to the fourth date. This brings us to August 9, A.D. 787. On this day Venus had the greatest eastern elongation.

Stela K from Quirigua contains the date 9-18-15-0-0 3 Ahau, 3 Yax equivalent to December 31, A.D. 816. On this day Mercury rise heliacally and the mean sun entered the ancient Hindu constellation Capricornus.

There is another date, 1 Oc, 18 Kayab got by deducting 10 Uinals and 10 Kins from the former date and equivalent to June 3, A.D. 816. On this day Saturn began to retrograde and the planets Mercury and Jupiter were in conjunction.

Altar G at Copan contains the date 10 Ahau, 8 Zac corresponding to the long count date 10-1-2-13-0. This date is equivalent to January 13, A.D. 864. This was a new moon day on which there was the circumstance of a solar eclipse and Mercury set heliacally on this day. There was a lunar eclipse the very next fortnight on January 27. These eclipses are represented in the Altar as will be evident from the following description: 'The Altar (G at Copan) represents a double headed feathered serpent. From the open jaws at both ends protrudes the upper half of a human figure '. This is evidently the Hindu demon Rahu *(Caput Draconis)* having a body consisting of *head and tail, the figure of a snake,* devouring and disgorging the sun and the moon, with Thompson's epoch the date is equivalent to August 2, A.D. 852 which was not a new or full moon day even. Mr. Thompson, however, considered the date 10 Ahau, 8 Zac to correspond to the long count date 9-18-10-0-0 equivalent to the Gregorian date August 20 (August 15 Jul.), A.D. 800 (' Civilization of the Mayas ', pp. 96-97). But this date again is not even a new or full moon day, nor any planetary phenomena occurred on this day. Mr. Thompson remarks: ' The connection with Venus is not very clear '. With Kreichgauer's epoch of the Mayan calendar, June 7, 2997 B.C. (' Anthropos ' for 1927) the above dates are equivalent to May 2, A.D. 969 and May 15, A.D. 917. But none of these days are even a new or full moon day.

Altar U from Copan contains the corrected date 3 Caban O Pop corresponding to the long count date 9-15-9-10-17 and equivalent to July 16, A.D. 752. This was a new moon day on which occurred a partial solar eclipse and on this day Venus attained her greatest brilliancy. A fortnight later on July 31 there occurred a total lunar eclipse visible from Mexico. Another fortnight later on August 14 there occurred a partial solar eclipse. The two solar eclipses one on each side of the total lunar eclipse are evidently represented on the Altar U. 'The front is carved in the form of a grotesque face, probably that of a conventionalized snake. On each side are represented serpents in whose jaws is seated in each case a human figure.' That new moons and eclipses are described in this altar will be evident from this: 'Then follow a number of other dates including at the top left hand corner of the side the date 6 Caban, 10 Mol,—the date most frequently recorded at Copan'. We have already discussed this date corresponding to the long count date 9-16-12-5-17 in connection with Altar Q from Copan. This date has already been found to be a new moon day (December 8. A.D. 774). Now the interval between this date (9-16-12-5-17) and 9-15-9-10-17 is one of 8180 days which we know is exactly 277 lunations. Thus the significance of this date 9-15-9-10-17, 6 Caban 10 Mol most frequently recorded at Copan is now evident. Mr. Thompson could not find the exact significance of the date and remarked, 'The exact significance of this date can only be surmised. Possibly it marked some important political event in the history of Copan.' Neither Kreichgauer's nor Thompson's epoch yields any eclipse, or new or full moons on the above dates.

Stele 1 from Copan contains the date 9-12-3-14-0 equivalent to August 30, A.D. 687. On this day Mercury set heliacally and the sun entered the ancient Hindu constellation Virgo. Two days previous to this date full moon occurred and three days later Jupiter rose heliacally. These are perhaps mentioned in the two or three other dates which are not stated by Mr. Thompson remarking that these are of little interest to the general public. The back of Stele 1 contains the above-mentioned date. But there are other dates also on this side which are not found mentioned by Mr. Thompson. We find that one fortnight before the full moon date, i.e. on August 13, A.D. 687 there occurred a partial solar eclipse. One fortnight before this latter date, i.e. on July 30, A.D. 68 there occurred a total lunar eclipse visible from Mexico, the middle of the eclipse occurring about midnight. Another fortnight before this latter date, i.e. on July 15, A.D. 687 there occurred a partial solar eclipse. These two solar eclipses one on each side of the total lunar eclipse of July 30 are evidently represented by the two heads of the sun god peeping forth from the jaws of the double headed snake: ' The front of Stele 1 represents a full length figure above whose head is an elaborate mask with the four leaf hieroglyph symbol of the sun ...... In the folded arms of the figure rests the body of a double-headed snake, from whose jaws peep forth two heads, which, judging by their filed teeth represent the sun god.' Compare with this the Hindu belief of the sun (*Pushan*) being toothless (*adantakah*).

Altar K from Copan contains the date 9-12-16-7-8 equivalent to February 11, A.D. 700. On this day planet Saturn set heliacally and full moon occurred two days previously on February 9.

Altar D from Copan opens with the date 13 Ahau 8 Zac. This date is equivalent to the long count date 9-19-6-4-0 and equivalent to January 22, A.D. 828. Only two days previously on January 20 there occurred a partial solar eclipse. One fortnight before this date on January 6, A.D. 828 there occurred a total lunar eclipse visible from Mexico. Another fortnight before this latter date, i.e. on December 22, A.D. 827 there was the circumstance of a solar eclipse. From the description of the carvings on the Altar it is clear that eclipses are being mentioned there: ' The front side is carved with a large double headed dragon monster. From one head protrudes a human face.' All this reminds one of the Hindu Rahu devouring and disgorging the sun and the moon. [In the above three dates from Copan a difference of two days has been observed from new or full moon days. If, however, the Julian day number 588465 of February 17, 3102 B.C. is applied in converting the dates—the resulting dates differ from new or full moon days by one day only.]

In the last picture of the lunar count in the Dresden Codex occurs the date 9-17-17-14-6, equivalent to January 15,

A.D. 800. This was a full moon day on which occurred a partial lunar eclipse and the two planets Jupiter and Saturn were in geocentric opposition.

The Stela 8 at Copan contains the date 9-17-12-6-2 9 Ik 15 Zip, equivalent to August 30, A.D. 794. This was a new moon day on which Jupiter was in conjunction with the sun.

Stela J at Quirigua opens with the date 9-16-5-0-0. This date is equivalent to September 19, A.D. 767. The autumnal equinox occurred on this day and the planets Mercury and Venus set heliacally in the east jointly on this day. ' In glyph C 4 of this Stele there is a secondary series of one uinal and one kin. Subtracting this the date 9-16-4-16-19 is reached', i.e. 21 days before the previous date, equivalent to August 29, A.D. 767 which curiously we know was a new moon day.

Altar 1, at Copan contains the date 9-16-11-0-5, equivalent to August 23, A.D. 773. This was a new moon day. Mercury set heliacally in the west three days before and Saturn set heliacally three days after this date.

A few other Mayan dates in a tabular form with their equivalents in Christian years after referring these to the epoch of the Kali Yuga era (February, 18.00, 3102 B.C. Julian day number 588466) with the calculated results of astronomical phenomena occurring on those days are presented below:

(1) 8-6-2-4-17=25th Aug., A.D. 173.	Sun's long. 150°. Jupiter stationary on the 26th and greatest eastern elongation of Mercury on the 27th. New moon on the 26th on which there was the circumstance of a solar eclipse. The next fortnight a lunar eclipse and another fortnight later a solar eclipse. Mars in opposition with sun of the 30th.

(2)	8-14-10-13-15= 21st Sept., A.D. 339.	Sun's mean long. 180° and autumnal equinox two days later. New moon on the 10th very nearly an eclipse. The next fortnight a lunar eclipse and another fortnight later a solar eclipse.
(3)	9-6-10-0-0=9th July, A.D. 575	Full moon day. Retrogradation of Jupiter begins on 13th.
(4)	9-9-0-0-0=19th July, A.D. 624	Sun's mean long. 210°. New moon on the 18th and greatest eastern elong. of Mercury on the 17th.
(5)	9-9-10-0-0=27th Aug., A.D. 634	New moon on the 29th. Jup. in opp. with sun on 30th.
(6)	9-13-17-12-10= 2nd Feb., A.D. 721	New moon on February 1.
(7)	9-15-0-0-0=28th Jan., A.D. 743	Hel. rising of Mercury in west. New moon and a solar eclipse on the 30th.
(8)	9-15-9-13-0=28th Aug., A.D., 752.	Venus and Jupiter in opp. in R.A. Greatest eastern elong. of Mercury on the 25th Full moon on the 29th.
(9)	9-16-10-0-0=23rd Aug., A.D. 772	Venus and Jupiter in conjunction. Greatest eastern elong. of Mercury on the 20th.
(10)	9-17-5-4-5=30th Aug., A.D. 787.	Hel. rising of Mars and greatest eastern elong. of Mercury on the 31st. A lunar eclipse two days later and a solar eclipse the next fortnight on Sept. 16. Mars and Saturn in opp. on Sept. 4.

4
The Origin and Development of Numerals

I

The Meaning and use of Numerals

"The numerals or the nine figures, with the device of places, to make them suffice for all numbers," says Bhaskara in his Vasana, "originate from the beneficent Creator of the Universe." Such was the great importance of the numerals in the daily life of human beings that the Hindus were more inclined to ascribe their origin to the Divinity than to human agencies. It might mean, of course, that the great antiquity of these numerals among the Hindus was beyond calculation, having been used at a period anterior to all existing records. This was also the practice with the Greeks, Egyptians and most other nations, with respect to more important inventions in the arts of life whose origin is lost in the remoteness of antiquity.

The chief use of numerals is for reckoning. The use of visible signs to represent numbers and aid reckoning is not only older than writing, but also older than the development of numerical language on the denary system. We count by tens because our ancestors counted on their fingers and named the numbers accordingly. So used, the fingers were with our ancestors really numerals, that is visible numerical signs; and in remote antiquity the practice of counting by these natural signs were in vogue in all classes of society.[1] In the later times of antiquity the finger symbols were developed into system capable of expressing all numbers below 10,000. The left hand was held up flat with the fingers together. The units from one to nine were expressed by various positions of the third, fourth and fifth fingers alone, one or more of these being either close on the palm or simply bent at the middle joint, according to the number meant. The thumb and index were thus left free to express the tens by a variety of relative positions, e.g., for thirty their points were brought together and stretched forward; for fifty the thumb was bent like Greek P and brought against the ball of the index. The same set of signs, if executed with the thumb and index of the right hand, meant hundreds instead of tens and the unit signs if performed on the right hand meant thousands[2]. The fingers served to express numbers, but to make a permanent note of numbers some kind of mark or tally was needed. Alphabet writing did not do away with the use of numerical symbols, which were more perspicuous and compendious than words written at length. But the letters of the alphabet themselves came to be used as numerals. One way of doing this was to use the initial letter of the name of a number as its sign.

Numerals, therefore, mean the instruments of reckoning. Needs of reckoning were felt with the very dawn of human consciousness, for expression of ideas required reckoning. Hence it is conceivable that reckoning came into force from the time human beings became articulate. Human literature, that is available at the present time as records of the oldest antiquity, bears testimony to this fact. The oldest records of Hindu literature, i.e., the Vedas, show that reckoning was of everyday necessity in practical life. On astronomical grounds some Sanskrit scholars hold that the oldest Vedic hymns date from 3,000 B.C., while some others put them as far back as

6,000 B.C., it is also an established fact that the Rg-veda is the oldest literary document of the Indo-European languages. There also we find in the famous Savitri hymn:

तिस्त्रो दृयावः सवितुर्द्वा उपस्था एका यमस्य भुवने विराषाट्।

There are *three* heavens; *two* are the laps of Savitri, *one* overcoming men, in the abode of Yama. (I, 35, 6).

or

अटौ व्यख्यत्ककुभः पृथिव्याम् श्री धन्व योजना सप्त सिन्धून्।

He has surveyed the *eight* peaks of the earth, the *three* waste lands, the leagues and the *seven* rivers (I,35, 8).

And also in the famous Purusa hymn:

सहस्रशीर्षा पुरुषः सहस्राक्षः सहस्त्रापात् स भूमिं विश्वतो वृत्वात्यतिष्ठद्दशाङ्गलम्।

Thousand-headed was Purusa, *thousand*-eyed, *thousand*-footed. He having covered the earth on al sides, extended beyond it the length of *ten* fingers. (X, 90, 1).

Here it is seen that fingers were not only used for measuring length but also for reckoning.

This system of reckoning by numerals has been adopted, with slight variations, not merely in all languages of the same class and origin, but likewise in many others which are radically different from them. If we proceed to the expressions of higher numbers, we find the same general law of their formation by the combination of names of the articulate numbers, with those of the nine digits.

Speaking of the antiquity of the numbers Dr. Peacock in his work on arithmetic observes:—"From the consideration that when a national literature whether oral or written, is so generally diffused as to form a standard, or a test of purity, which while it enforces a legitimate character upon all existing terms, watches over the introduction of all others with extreme jealousy; from this consideration alone, independently of other evidence, we should be inclined to assign to the Sanskrit terms for high numbers, and consequently, to their system of numeration upon which they are founded, an antiquity at least as great as their most ancient literary monuments; as the arbitrary impositions of so many new names for the most part independent of each other, and in numbers, also, so much greater than could possibly be required for any ordinary application of them, would be a circumstance entirely without example in any language which had already acquired a settled and generally recognised character."

This much is sufficient to prove the antiquity of numerals. Of their use in our daily life it is needless to add more than what has been said by Bhaskara that their importance has been such as to lead ancient people to think them to have originated from the Creator himself. We have seen what the general meaning of numerals was and has been and in their early conception the fingers played the principal part. So much useful have been the numerals with people all over the world from the very ancient times and so much familiar are we with them, that it is difficult to ascertain correctly their origin and more difficult to find the place of their birth. There are, moreover, such scanty materials to fall upon that the task of determining their origin becomes almost improbable, if not impossible. The meaning and use of numerals are so self evident that we are often led to believe that Bhaskara was right when he attributed the origin to divine agency and not to any human authorship.

II

The Hindu-Arabic Question

In histories of the mathematical sciences it has been usual to trace the knowledge of arithmetic to the Arabs, and the numerals are distinguished by the symbols termed Arabic.

Dr. Peacock, in his work on arithmetic, observes that there is nothing in the Greek notation which in the slightest degree resembles the present numerals, and nothing in the object proposed in the researches of Archimedes and Appolonius which could naturally lead to their invention.

The idea that the numerals are Arabic in origin is not an old one. The Mediaeval and Renaissance writers generally recognised them as Indian, and many of them expressly stated that they were of Hindu origin.[3] Others were of opinion that the numerals were probably invented by the Chaldeans or the Jews because they increased in value from right to left. Robert Recorde (c. 1542), England's earliest arithmetical text-book writer, probably referred to this general idea of notation when he wrote: "In that thing all men do agree, that the Chaldays, whiche fyrste invented thys arte, did set these figures as they set all their letters for they wryte backwards as you tearme it, and so do they reade. And that may appeare in all Hebrewe, Chaldaye and Arabike books ..... whereas the Greekes, Latines, and all nations of Europe, do wryte and reade from the left hand towarde the ryghte."[4] There were others also, among whom were such influential writers as Tartaglia in Italy and Kobel in Germany", who asserted the Arabic origin of the numerals on the ground of their order being from right to left. This order of writing probably led Mr. G. R. Kaye to think that the numerals are of Arabic origin. "Sanskrit and kindred scripts" says Mr. G. R. Kaye "are, and have been, for centuries, written from left to right, while the Arabic family of scripts are written from right to left. It would be natural to expect number, words and symbols to be affected by the mode and direction of writing."[5] But how is it then that the Arabs themselves never laid claim to the invention of numerals, and always recognised their indebtedness to the Hindus both for the numeral values and for the distinguishing feature of place value? The foremost among the Arab writers on mathematical classics, Muhammad, the son of Moses, from

Khowarezm, or more after the manner of the Arab, Muhammad ibn Musa al-khowarezmi, a man of great learning and one to whom the world is much indebted for its present knowledge of Algebra and Arithmetic, distinctly stated that the numerals were of Hindu origin. Even the later Arab writers, down to the present day, use the phrase 'ilm hindi,' "Indian science," for arithmetic as also the adjective 'hindi' alone. Probably the most striking testimony from Arabic sources is that given by the Arabic traveller and scholar Muhammad ibn Ahmed, Abu 'l-Rihan al-Biruni (973-1048 A.D.) who spent many years in Hindustan. He states[6] explicitly that the Hindus of his time did not use the letters of their alphabet for numerical notation as the Arabs did, and that the numerical signs called *anka*[7] had different shapes in various parts of India, as was the case with letters. In his Chronology of Ancient Nations he gives the sum of a geometric progression and shows, how, in order to avoid any possibility of error, the number may be expressed in three different systems:—with Indian symbols, in sexagesimal notation, and by an alphabet system. He also speaks of "179, 876, 755, expressed in Indian ciphers," thus again attributing these forms to Hindu sources.[8] Preceding Al-Biruni, another eminent Arabic writer of the tenth century, Motahhar ibn Tahir in his "Book of the Creation and of History" gave as a curiosity, in Indian Nagari symbols, a large number asserted by the people of India to represent the duration of the world. In Motahhar's time the present Arabic symbols had not yet come into use, and that the Indian symbols, though known to scholars, were not current.[9] Because of this fact Motahhar found something extraordinary in the appearance of the number he cited. Another Arabic writer named Al-Kindi (800-870 A.D.) wrote five books on arithmetic and four books on the use of the Indian method of reckoning. We are therefore: forced to the conclusion that the Arabs from the early ninth century on fully recognised the Hindu origin of the numerals.

But Mr. G.R. Kaye does not think the above evidence sufficient to prove the Hindu origin of the numerals. He says, "In the early stages of any language we generally find that the smaller elements of the higher numbers are expressed first. Thus we have *two and hund seofontig* in Alfred's Chronicle for the modern 'seventy-two' and the Germans still keep to this fashion much more than we do. Such examples as fifteen, trydaca (त्रयोदश) are found in many languages." He further adds, "The order in which we write our numbers is contrary to the nature of our script and has been imposed on us really by a people with a right to left script. This conclusion, if generally acknowledged as correct, would appear to dispose of the question as to the notation in use being of Indian origin." As an illustration of his theory Mr. Kaye says that the date "Saka samvat, 867" (A.D. 945 or 946) is given by " giri-rasa-vasu" meaning "the mountains" (seven), "the flavours" (six), and the gods "Vasu" who were eight in number. But here Mr. Kaye is misinformed. In reading the date these were read from right to left. The period of invention of this system is uncertain. The first trace seems to be in the Srautasutra of Katyayana and Latyayana. It has been asserted by W. Brennand that Aryabhata was also familiar with this system.[10] But it was certainly known to Varaha-Mihira (d. 587), for he makes use of this system in his Brhat-Samhita. In Chapter VIII, sloka 10 of this work in the course of calculation of a particular time he mentions *bibhajayecchunyasaragaramaih,* ("should divide by 3750," Sunya—zero, Sara—arrow—five, aga—mountain—seven, Rama—3, as there were three Ramas.—Rama, Parasurama and Balarma)'[11] Here the reckoning is made from right to left. So that the objection of Mr. G. R. Kaye to the numerals being of Hindu origin on the ground of the order not being from right to left does not stand at all. Further, Al-Biruni remarks that the Hindus writing in slokas changed their orders, though they generally reckoned from right to left, for otherwise, the meter would have been disturbed. Apart from this word-and-letter system there was the old *Kharosthi* dialect prevalent in the North-West of India, the characters of which were written from right to left instead of from left to right. Of this, probably, Mr. G.R. Kaye took scanty notice, and hence he was so much perplexed to think of the order. For, this *Kharosthi* script affects the Hindu-Arabic question vitally.

Next to the finger calculation came in vogue the abacus, a mechanical contrivance, for reckoning. It had been, introduced very early for keeping number of different denominators apart. Abacus literally means a wooden plank. It was a table with compartments or columns for counters, each column representing a different value to be given to a counter placed on it. This might be used either for concrete arithmetic—say with columns for rupees, annas, pies; or for abstract reckoning. An old Greek abacus found at Salamis has columns which, taken from right to left, give a counter the value of 1, 10, 100, 1000 drachms, and finally of one talent (6000 drachms) respectively. An abacus, on the decimal system might be ruled on a paper or on a board, strewn with fine sand, and was then a first step to the decimal system. Two important steps, however, were still lacking; the first was to use instead of counters distinctive marks (ciphers) for the digits from one to nine; the second and the most important was to get a sign for zero, so that the column might be dispensed with and denomination of each cipher seen at once by counting the number of digits following it. These two steps being taken, the present system of the so called Arabic numerals and the possibility of modern arithmetic are reached; but the invention of the cipher and zero came slowly.

Sir E.C. Bayley says, "It need hardly be said that the use of the abacus is still common in every village bazar in India, and has been universal apparently from time immemorial".[12] Burnell tells us that the Indian abacus was by using heaps of couries for the numbers, the number of these shells being equal to that of the number expressed, the cipher being a blank space. Thus "∴ 11 = 303; ∴1∴ = 33."[13] Mr. S. B. Dikshit writes: "Hindu astrologers use a wooden plank, which they cover with dust; this plank is called *pati;* hence arithmetic is called Patigarita by Bhaskaracarya and others." [14] Bayley also maintains this view. The mention of this abacus is necessary inasmuch as it leads to the invention of cipher or zero without which the system of numerals was not complete. Taylor, Woepcke, Bayley, Burnell and others derive the Sanskrit numeral words that signify zero from, the use of the abacus. These terms, they say, indicate the space not-filled up by a counter. All these terms indicate emptiness or the sky, ether, etc. The term cipher is derived from *sunya* through the Arabic *syfr.* "In Sanskrit" wrote Taylor in 1816; "the word Sunya signifies a circle, cipher or vacuity; and the Arabs on receiving the numerical notations from India translated it by the word *syfr,* which in their language also means emptiness, vacuity or nothing.[15] The origin of the abacus has a special importance in the history of the origin of the numerals. But where did this abacus originate? Bayley and Burnell say that it was frequently used in India. How is it then that there is no mention of this in any Sanskrit literature?

Previous to the advent of the oriental scholars it was the consensus of opinion in Europe that the numerals were of Arabic origin and they were generally termed Arabic numerals. We have already mentioned that such influential writers as Tartaglia in Italy and Kobel in Germany asserted the Arabic origin of the numerals. But in later days the laborious researches of the oriental scholars led most of them to believe the numerals to be of Hindu origin. S. Corrington in his history of arithmetic says, "In the section called Algorithims of Brahmagupta's mathematics we have undoubtedly the numeration and notation of the Hindu system given and explained. In the twelfth century Bhaskara composed a fuller and more valuable work on arithmetic, and undoubtedly there was a race of scholars (between Brahmagupta and Bhaskara) during the intervening centuries to whom was due the maintenance and extension of Hindu learning. Thus the mathematical writings of the Hindus became known to the Arabians and specially the wondrous system of notation having nine digits and a cipher with device of a place." Rodet shows that Aryabhatta's rule for the extraction of square root implied a knowledge of the value of position:—

भागं हरेद् अवर्गान् नित्यं द्विगुणेन वर्गमूलेन।

वर्गाद् वर्गे शुद्ध लब्धं स्थानान्तरे मूलम्।।

"Always divide the part that is not square by twice the root of the square, after having subtracted from this squared part the square of the root; the quotient is the root of the next term."

But this implication, says Mr. G.R. Kaye, is no proof of the origin of the numerals among the Hindus. Of the oriental writers, only a very few, Mr. G.R. Kaye and others, doubt the Hindu origin of the numerals, but at the same time cannot prove that they owe their origin to the Arabs.[16]

Apart from the assertion of the oriental scholars who belong mostly to the western countries, we have already cited the evidence of the Arabs themselves. Al-Biruni, the great scholar and traveller, conclusively says, "numeral signs which we use are derived from the forms of the Hindu signs. The Hindus use the numeral signs in arithmetic in the same way as we do. I have composed a treatise showing how far the Hindus are ahead of us in the subject." Even Mr. S. Khoda Bux, an eminent Islamic scholar, in a recent article on "Literary and scientific activities under the Caliphate," says, "The Arabs in the IX century borrowed from the Indians their decimal system, numerals and arithmetic." In this chapter we have opened the question of the origin of Hindu-Arabic numerals, a solution of which will be attempted by an examination of the materials hitherto available.

III

The Origin and Development of Hindu Numerals with the Invention of Zero and the Place Value

"It is known," says Sir E. C. Bayley, "that while in European parlance the modern numerals have been termed 'Arabic' yet that they differ in several important points from those used in Arabic writing; moreover Arab writers emphatically declare these last to be an Indian invention. Still, the chain of descent of either form from an Indian source has never been satisfactorily elucidated. And while no doubt the modern system of using, for all purposes of notation and calculation, nine unit figures and a zero, arranged in decimal order, is apparently indigenous in India, it has not yet been distinctly shown how this simplified form was eliminated from the complicated system of notation which was in ancient times used in India."[17]

It is with this unsatisfactory material that we have to deal in searching for the early history of the Hindu Arabic numerals, the probability being that writing was not introduced into India before the close of the fourth century B.C. and literature existing only in spoken form prior to that period.

The whole question of the Hindu numerals therefore divides itself into two parts, *vis.*,

(1) The origin of the ancient Indian system of numerals;

(2) The simplification of this system, by the rejection of all the signs except those for the nine units, and by the invention of the zero.

The early Hindu numerals may be classified into three great groups, (i) the Kharosthi, (2) the Brahmi and (3) the word and letter forms.

The Kharosthi numerals are found in inscriptions formerly known as Bactrian, Indo-Bactrian, and Aryan, and appearing in ancient Gandhara, now eastern Afganisthan and northern Punjab. The alphabet of the language is found in inscriptions dating from the fourth century B.C. to the third century A.D., and from the fact that the words are written from right to left it is assumed to be of Semitic origin. Not until the time of king Asoka, in the third century B.C., do the numerals appear in any inscriptions, thus far discovered; and then only in the primitive forms of marks. These Asoka inscriptions are found in widely separated parts of India, often on columns, and are in the various vernaculars that were familiar to the people. Two are in the Kharosthi characters, and the rest in some form of Brahmi. In the Kharosthi inscriptions only four numerals have been found and these are merely vertical marks, for one, two, four and five, thus,—1, 11, 1111, 11111. In the so-called Saka inscriptions, possibly of the first century B.C., more numerals are found, and in more highly developed form, the right-to-left system appearing, together with evidences of three different scales of counting four, ten and twenty. The numerals of this period are as follows,—

1 2 3 4 5 6 7 8 9

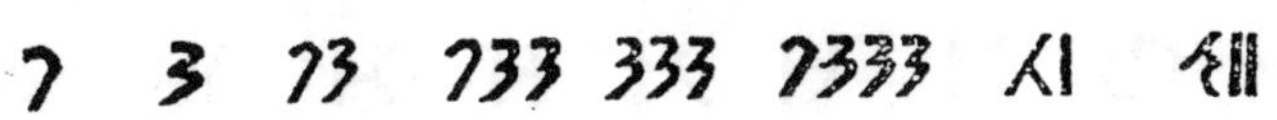

10 20 30 50 60 70 100 200

In this system there are several noteworthy points. In the first place, it is probably not as early as that shown in the Nana Ghat forms:

1 2 3 4 6 7 9 10 10 20 60

although the inscriptions themselves at Nana Ghat are later than those of the Asoka period.

It is not in the Kharosthi numerals, as we see, that we can hope to find the origin of those used by us, and we turn to the second of the Indian types, the Brahmi characters. The alphabet attributed to Brahman is the oldest of the several known in India, and was used from the earliest historic times. There are various theories of its origin. The numerals are not as old as the alphabet, or at least, they have not been found in inscriptions earlier than those in which the edicts of Asoka appear.

The following numerals appear in the Asoka edicts:

1 2 4 6 50 200 200 200

Here we notice that the "one" and "six" of the Brahmi numerals resemble the same of modern numerals. These fragments from the "third century, crude and unsatisfactory as they are, are surely forerunners of the present system.[18]

The next use of numerals is found in the Nana Ghat inscriptions, supposed to date from the early part of the third century B.C.

1 2 3 4 4 6 7 9 10 10 10 20

60 80 100 100 100 200 200 400 700

1000 1000 4000 6000 10000 20000

These are signs for units, tens and hundreds.[19] Along with these we consider that the "cave" numerals found in certain other cave inscriptions were employed during the first centuries of the Christian era. In the Nasik cave inscriptions, for example, the following forms are found.[20]

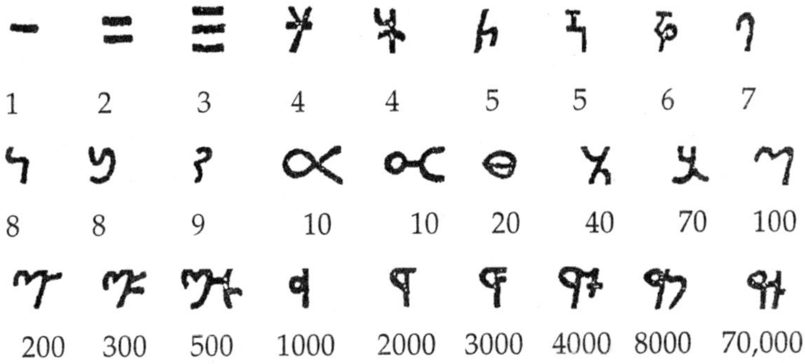

In all these systems we notice that no zero appears and as a matter of fact none existed in any of those cases cited. It was therefore impossible to have any place value, and the numbers like twenty, thirty and other multiples of ten, one hundred and so on, required separate symbols. Therefore a large number of symbols had to be used.

These are the earliest forms of Hindu numerals. No one can precisely say how these forms came to be used. The origin of some is evident, but probably the origin of most of them will never be known. Some say that they are derived from the alphabet of the Bactrian civilisation of Eastern Persia and some again find a possibility of Chinese influence in certain of the early numerals of the Hindus.[21] But these are mere guess works and cannot be relied upon.

We have thus far spoken of the Kharosthi and Brahmi numerals; now we shall mention the third type, *viz.*, the word and letter forms, with which the perfected system of the Hindus with the symbol zero is closely connected. This system came to be used as early as the sixth century of the Christian era, if not earlier. For example, Brahmagupta says in his *Brahmasphutasiddhanta,* "If you want to write one, express it by everything unique, i.e. one by adi (beginning), sasin, indu (moon), sita, dhara, urvara, bhu (earth), Brahman, Pitamaha, rasmi, etc., two by yamala, yama (twins), kara (hands) netra, aksi, clarsana, locana (eyes), Asvins' (the two brothers), etc.;

three by Rama (the three Ramas), guna (virtue); four by sagara (ocean); five by visaya (senses), sara (arrows), etc.; six by rasa (flavour), rtu (seasons), etc., seven by aga (mountain), and so on; ten by asa, dis (quarter) kendu, Ravanasura (because he had ten heads); zero by sunya (void), ambara, akasa (heaven, space), purna (circle), etc."

As an example of this system, it has already been said that the date "Saka Samvat, 867" (A.D. 945 or 946) was given by "giri-rasa-vasu," meaning the mountains (seven), the flavours (six), and gods "vasu" (who were eight in number) i.e. 867, the words being read from right to left. The period of invention of this system is uncertain. But as we have already remarked that the first trace seems to be in the Srautasutra of Katyayana and Latyayana and it is quite certain that Varahamihira (d. 587) used it in the Brhat-samhita (chap. VIII, 20). In the Pancasiddhantika also Varahamihira citing details laid down by Latacarya in a work in which he explained the Romaka and Paulisa Siddhantas tells us that for some calculative purposes a certain moment was assumed to be fixed (as regards the year)

सप्ताश्विवेदसंख्य' शककालमपास्य चैत्रशुल्कादौ

i.e. by deducting the Saka-year having the number seven, the two Asvins, the four Vedas i.e. 427 or 506 A.D. at the end of the bright half of the Caitra month.

This method of calculation came down even to the time of Bhaskra (1150 A.D.) who while giving in his Siddhantasiromani details about this birth and parentage says:

रसगुण–पूर्ण महीसमशकनृपसमयेऽभवन्ममोत्पत्तिः
रसगुणवर्षेण मया सिद्धान्तशिरोमणी रचितः।

i.e. I was born in the Saka-year 1036 (rasa—flavour—six, guna— virtue—three, purna—circle—zero, mahi—earth—one) and wrote *Siddhanta-siromani* in my 36th year (rasa—six, guna—three), (the words being read from right to left).

Next in importance is the alphabetic system invented by Aryabhata to express numbers. Aryabhata's rule giving his alphabetic system of expressing numbers is as follows:—

वर्गाक्षराणि वर्गेऽवर्गेऽवर्गाक्षराणि कात् ङ्मौ यः।
खद्विनवके स्वरा नव वर्गेऽवर्गे नवान्त्यवर्गे वा।।

It may be translated thus: Varga consonants from क (onwards) [should be used] in varga places (i.e., places corresponding to the varga or square units 1, 10^2, 10^4 &c.) and avarga consonants in avarga places i.e. places corresponding to the avarga or non-square units 10, 10^5, 10^5, &c.).; य stands for 30 (lit., 5 and 25). Vowels should be used in eighteen places, nine vowels (with distinctly different sounds) in varga places as well as in the (corresponding) avarga places. Those nine vowels should be used in higher places in a similar manner.

The above translation differs from that given by the late Dr. Fleet (J.R.A.S., 1911, p. 115). He writes: "The concluding words 'navantyavarge va' or 'in the square immediately following the nine' that is 'in the tenth square place' are enigmatical. They seem to indicate a ninteenth place (the number belonging to which the British trillion, would be square of the Vrnda No. 10) and nothing after it" (ibid, p. 120). The alleged engimatical nature of the world disappears when it is noticed that 'navantyavarge' is not a single compound word (समासवद्ध पद) but has been formed by joining the words 'nava' and 'antyavarge' according to the rules of svarasandhi or conjunction of vowels. 'antya' means' 'following' or 'which comes after' and "varga" means a group of the same class.' So the compound word 'antyavarge' means 'in the following group of the same class.' As a group of eighteen places has already been spoken of in the verse, the word 'antyavarge' can only mean 'in the following group of eighteen places.' In Sanskrit the word va (वा) is used in the sense of 'or' as well as in the sense of 'and.' Here it cannot mean 'or'; it must have been used in the sense of 'and.' The word 'va' has also the force of similarity (सादृश्य). Hence the suggestion is that the number of place is use limited and that nine vowels are to be used, as explained before, in each group of eighteen places.[22] Fleet translates the second line of the above verse thus: "the nine vowels (are used) in the two nines of spaces square (and) not square, or in the square immediately following the nine." How can nine vowels be used in 'the square following the nine' by which expression Fleet means the ninteenth place? Only one vowel is to be used in each place. This is sufficient to prove that Fleet's translation is incorrect.

It has been suggested[23] that Aryabhata's system of expressing numbers has been derived from the alphabetic notation of the Greeks. The reason for the above suggestion has been given,[24] by the late Dr. Fleet in the following words: "Knowing the Greek source of the "greater part of the astronomy, etc., which we have in the Aryabhatiya and subsequent works, we naturally think of the possibility of a similar origin for this system of numeration." Does the question of borrowing at all arise in connection with the original matter which is included with admittedly borrowed matter in the work of a western savant? In giving his value of π Aryabhata has used, not the Greek myriad or any of its corrupted forms, but the word 'ayuta' which is equivalent to the Greek myriad and has been in use in India at least since the time of the Vedas[25] (i.e. long before the Greek civilisation came into being). Yet it is asserted[26] that the way in which Aryabhata has expressed his value of π points to a Greek source on the alleged ground that the Greeks alone of all peoples use the myriad as their unit of the second order. But has any European critic ever thought of the possibility of the Greek myriad having been borrowed from India and of

Pythagoras's calling ten, hundred, etc., units of the second, third, etc., course after the Vedas which contain the earliest record of the use of a strictly decimal system of numeration, the different units being 1, 10, and higher powers of 10?"[27] The renowned German mathematican Euler included both original and borrowed matter in his writings. When "he employed practically the same method of solution"[28] of the so-called Pellian equation as was given some centuries before by the Hindu mathematicians, did anyone naturally think of the possibility of Euler's indebtedness to India for the solution? On the contrary, an attempt has been made[29] to belittle the importance of the equation in order to prove that the Hindus achieved nothing of importance in the field of mathematics.

"Priority of statement of a proposition does not necessarily imply its discovery." Previous use alone of an alphabetic notation by the Greeks, should not, therefore, lead us to trace Aryabhata's alphabetic system to a Greek source.

If the second verse of the Ganitapada (i.e. chapter II) of the Aryabhatiya be read with the corresponding verses of Mahavira and Bhaskara, it will be seen that it contains an exposition of the modern place-value decimal notation.[30] No one has hitherto claimed that the modern place-value notation was known to the Greeks or to any other non-Hindu people before the sixth century.[31] It cannot, therefore, be said that Aryabhata's rule explaining the modern notation was borrowed from the Greeks.

Now, the only point of similarity between the Greek and Aryabhatian alphabetic systems is that the first nine letters of the alphabet denote the first nine numbers in each case. Aryabhata's system differs from the Greek system in every other particular. The principal points of difference may be stated as follows:

(a) In the Greek system the second group of nine letters denoted the first nine multiples of ten and the third group of nine letters the first nine multiples of hundred. To express multiples of higher powers of ten, strokes or dots were used. Each stroke or dot indicated multiplication by 1,000. Fleet says[32] that Aryabhata's system is certainly not an adaptation of this system of the Greeks, but that Aryabhata derived his inspiration from another Greek system of expressing large numbers in which myriads used to be expressed by means of two letter numerals, viz., (i) a symbol M for a myriad and (ii) the already adopted symbol for the number indicating the multiple. In Indian Kharosthi and later Brahmi notations hundreds used to be expressed in this way. But in Aryabhata's system each multiple of a power of ten was denoted by a single consonant-numeral combined with a vowel-sign. Fleet is right in holding[33] that in Aryabhata's scheme the vowels had no numerical values in themselves but marked the places to which the consonants, etc. were to be referred. If they had any numerical value, they could be used to express component parts of a number where no confusion was likely to arise. For example, (दश); could be used to denote 170. But vowels as such were never used by Aryabhata; and vowel-signs cannot stand by themselves.

(b) As has been already stated, Aryabhata's vowel-signs are not numerical symbols but indicate places which the consonant-numerals occupy. Hence Aryabhata expresses numbers by means of consonant-numerals and as many places indicating vowel-signs. But the Greek system exclusively employs letter numerals only. Place-indicating signs are conspicuous by absence in the Greek system.

(c) Unlike the Greek system and the old and modern Indian systems[34] of notation and Indian way of speaking numbers, Aryabhata's system (i) recognises component parts of a number, which are higher multiples than the ninth of even powers of ten, (ii) makes no provision for expressing the first two[35] multiples of the old powers of ten as such, and (iii) admits of a number being expressed in more ways than one. For example, (i) मि (mi) stands for twenty-five hundreds; (ii) two thousand cannot be expressed as such; it should be regarded as twenty hundreds before it can be expressed by नि (ni); and (iii) forty-five can be expressed by मन, भय, बफ, यण or रड.

(d) The Greek alphabetic system was an arithmetical notation, i.e., it was used in performing arithmetical operations. But Aryabhata's system was not so. As an arithmetical notation it has many and grave defects and is, therefore, useless. Hence it has not been given any place in the arithmetical portion of the Aryabhatiya. Arithmetical operation could be more easily and rapidly performed with the previous Indian notations.[36] The only merit of Aryabhata's system is its conciseness and it has been devised chiefly to secure brevity of the rules composed in verse.

Aryabhata[37] had undoubtedly been a student of Sanskrit grammar and metrics. In Sanskrit grammar single letters are used for two different purposes, viz. (i) as a suffix (प्रत्यय, e.g., क, ट, etc.) and (ii) as a संज्ञा or name for something to which frequent reference has to be made (e.g. कु, घि दि, etc.). In Pingala's manual of metrics single consonants (e.g. भ, स, ज, etc.) have been used for the second purpose. To be included in metrical composition numbers must, of necessity, be expressed by word-numerals or letter-numerals. The study of Sanskrit grammar and metrics seems to have led the mathematical genius of Aryabhata to use letters of the Devanagri alphabet for the sake of brevity, as it afterwards led the well-known grammarian Yopadeva to use these letters in shortening the Sanskrit grammar. The vowels were not suitable for this purpose as they often disappear and merge into unrecognisable forms owing to conjunction (सन्धि) which is an essential feature of Sanskrit. Aryabhata had, therefore, no other alternative than to use consonants to express numbers. The modern place-value notation was known to Aryabhata who classified the places as varga and a-varga. Most probably phonetic resemblance was responsible for the rule "varga consonants should be placed in varga places and a-varga consonants in a-varga places." To indicate the varga or a-varga place occupied by a consonant nothing could be more convenient than a vowel-sign. Hence vowel-signs have been used as place-indicators. Aryabhata names the first ten places (ekam, dasa, etc. up to vrndam) only. The first ten vowels were perhaps intended to indicate these places. Here some difficulty presented itself. Of the two vowels अ, आ the second is a long अ. The vowels of each of the remaining four pairs have similar sounds, the first vowel being short and the second long. In books on Sanskrit grammar the vowels constituting each of the five pairs are called equal (समान) vowels. Hence Aryabhata seems to have overlooked the distinction between long and short vowels and made the rule that

अ (or आ) should indicate the places of the units . . . . 10^0, 10^1

इ (or ई) .... 10^2, 10^3

उ (or ऊ) .... 10^4, 10^5

ऋ (or ॠ) .... 10^6, 10^7

लृ (or ॡ) .... 10^8, 70^9

Thus, each of the five vowels, अ, इ, उ, ऋ, लृ long or short, was assigned to a varga place and the next higher a-varga place. This could not result in confusion, as varga places were to be occupied by varga consonants only and a-varga places by a-varga consonants only. Then by the principle of analogy—a principle which is responsible for two serious mistakes[38] made by Aryabhata—he also assigned the remaining four vowels ऐ, ग्रो, ग्रौ (which have distinctly different sounds), each to two consecutive places. Hence the rule "vowels should be placed in eighteen places, nine vowels in varga places and the same nine vowels in a-varga places." As by his rule, ञ and न denote the first and second multiple of 10, he used; य, र etc., up to ह for the 3rd, 4th, etc. up to the 10th multiple of 10. It should be noted here that only when the consonants; य, र, ल, व etc. are each associated with the vowel ग्र they denote 30, 40, 50, 60, etc. Thus, यि means; य (or 3) put in the thousandth place (the a-varga place assigned to इ). If; य stood for 30 यि would denote 30 tens or 300. Fleet is not right in holding that in Aryabhata's scheme consonants have no numerical value in themselves. The metre shows that this reading is correct. If he were right, ङ् in the rule "ङ् मौयः could not stand for 5.[39]

Mention should also be made, in this connection, of a curious system of alphabetic numerals that sprang up in Southern India. In this we have the numerals represented by the letters as given in the following table: —

1	2	3	4	5	6	7	8	9	0
क	ख	ग	घ	ङ	च	छ	ज	झ	ञ
ट	ठ	ड	ढ	ण	त	थ	द	ध	न
प	फ	ब	भ	म					
य	र	ल	व	श्	ष	म	ह		

For example, the word

2 3 1 5 6 5 1

Kha gant yan me sa ma pa (खगन्तयान्मेषमापः) has the value 1,565,132, reading from right to left.[40] This is given in a commentary on the Rigveda, representing the number of days that had elapsed from the beginning of the Kaliyuga. Burnell states that this system is even now in use for remembering rules to calculate horoscopes, and for astronomical tables.[41]

But these systems without the zero were impracticable to designate the tens, hundreds, and other units of higher order by the same symbols used for the units from one to nine. There was therefore no possibility of place value without some further improvement. So the Nana Ghat symbols required the writing of "thousand seven twenty-four" almost like T 7, tw, 4 in modern symbols, instead of 7024, in which the seven of the thousands, the two of the tens (concealed in the word twenty, being originally "twain of tens," the 'ty' signifying ten), and the four of the units are given as spoken and the order of the unit is given by the place. The system required the zero for its perfection. It is possible that one of the forms of ancient abacus suggested to some Hindu mathematician the use of a symbol to stand for the vacant line when the counters were removed. There were however different names for the higher orders of the Hindu numerals. We have in Bhaskara's Lilavati:

Ekadasasatasahasrayutalaksaprayutakotayah kramasah
Arbudam abjam kharvanikharvamahapadmasankavas tasmat
Jaladhiscantyam madhyam parardham iti dasagunottaram samjnah
Samkhyayah sthananam vyavaharatham krtah purvaih.[42]

I.e., older mathematicians have invented the names eka, dasa, sata, sahasra, ayuta, laksa, prayuta, koti, arbuda, abja, kharva, nikharva, mahapadma, sanku, jaladhi, antya, madhya, paradha, and such multiples of ten for indicating the place values of numbers. Indeed this may have necessitated the introduction of a word to signify a vacant place or lacking unit, with the ultimate introduction of a zero symbol for this world.

To enable us to appreciate the force of this argument a large number, 8,443,682,155, may be considered as the Hindus wrote and read it, and then by way of contrast as the Greeks and the Arabs would have read it.[43]

Modern American reading:—8 billion, 443 million, 682 thousand, 155.

Hindu:—8 padmas, 4 arbudas, 4 kotis, 3 prayutas, 6 laksas, 8 ayutas, 2 sahasras, 1 sata, 5 dasa, 5.

Arabic and early German:— eight thousand, thousand thousand and four hundred thousand thousand, and forty-three thousand thousand, and six hundred thousand, and eighty-two thousand and one hundred fifty five (or five and fifty).

Greek:—eighty-four myriads of myriads and four thousand three hundred sixty-eight myriads and two thousand and one hundred fifty-five.

The reading of numbers of this kind shows that the notation adopted by the Hindus tended to bring out the place idea. No other language than the Sanskrit has made such consistent application, in numeration, of the decimal system of numbers; nor did any other ancient people carry the numeration as far as the Hindus did.

When the ankapalli (the nine ciphers were called anka), the decimal place system of writing numbers, was perfected, the tenth symbol was called sunyabindu, generally shortened to sunya (the void). It is generally thought that this sunya as a symbol was used about 500 A.D.; but it is certain that in the sixth century the use was frequent as we have seen that Varahamihira in his Brhat-samhita uses sunya while speaking of numerals. Varahamihira used frequently the word system with place value in his Pancasiddhantika.[44] At the opening of the next century (c. 620 A.D.) Bana wrote of Subandhu's

Vasavadatta as a celebrated work and mentioned that the stars dotting the sky were there compared with ciphers. Traces of the numeral words with place value, therefore, are found very early in India. Bühler[45] gives the copper-plate Gurjara inscription of Cedi-samvat 346 (594 A.D.) as the oldest epigraphical use of the numerals "in which the symbols correspond to the alphabet numerals of the period and the place." Vincent Smith[46] quotes a stone inscription of 815 A.D., dated samvat 872. So F. Kielhorn in the Epigraphia Indica[47] gives a Pathari-pillar inscription of Parabala, dated Vikrama-samvat 917, which corresponds to 861 A.D. and refers also to another copper-plate inscription dated Vikrama-samvat 813 (756 A.D.). The inscription quoted by V. A. Smith above is that given by Dr. D. R. Bhandarkar (Epigraphia Indica, vol. i, pp. 193, 198) and another is given by the same writer as of date Saka-samvat 715 (798 A.D.), being incised on a pilaster. Kielhorn also gives two copper-plate inscriptions of the time of Mahendrapala of Kanauj, Valabhi-samvat 574 (893 A.D.) and Vikrama-samvat 956 (899 A.D.). That there should be any inscriptions of date as early even as 750 A.D., would tend to show that the system was at least a century older. "That the system now in use by all civilised nations is of Hindu origin cannot be doubted; no other nation has any claim upon its discovery, especially since the references to the origin of the system, which are found in the nations of Western Asia, point unanimously towards India."[48]

The testimony and opinions of men like Bühler, Kielhorn, V. A. Smith, Bhandarkar, and Thibaut are entitled to the most serious consideration. As authorities on ancient Indian epigraphy and cultural history no other rank higher, their work is accepted by Indian scholars the world over, and their conclusion as to the rise and development of the system with a place value—that it took place in India as early as the sixth century A.D.—must stand unless new evidence of great weight can be submitted to the contrary.[49]

The improved Hindu system with a place value would never have dominated the numeral system of the western world, unless there had been the symbol zero. "The earliest undoubted occurrence of a zero in India is an inscription at Gwalior, dated Samvat 933 (876 A.D.), where 50 garlands are mentioned (line 20), 50 is wrtten as घ॰, 270 (line 4) is written as 170.[50] Apart from the appearance of zero in early inscriptions, there are indications of the Hindu origin of the symbol in the special treatment of the concept zero in the early works on arithmetic. Brahmagupta in the early part of the seventh century gives in his arithmetic a distinct treatment of the properties of zero.[50] A still more scientific treatment is given by Bhaskara. Even in the most recently discovered work on ancient Indian Mathematical lore, the Ganita-sarasamgraha of Mahaviracarya (c. 830 A.D.), there is a discussion on the calculation with zero.[51] In the first chapter of this book we find the word and letter system of numerals, though there is no indication of the numerals with the place value. But in the second chapter there is a long discussion on the sunyabindu.

It is not till the time of Bhaskara (1150 A.D.) that we find traces in writing of a perfected system of numerals with the place value of zero. From that time the modern Sanskrit numerals

१	२	३	४	५	६	७	८	९	०
1	2	3	4	5	6	7	8	9	0

came into use.

But Bhaskara mentions the names of a number of works in the concluding chapter of his Bija-ganita from which he gathered materials for his work: —

"As the treatises of Brahmagupta, Sridhara, and Padmanabha are too diffuse, I have compressed the substance of those works in a well-seasoned compendium for the benefit of learners." (verse 2).

Colebrooke was fortunate enough to secure a copy of Sridhara's arithmetic and there he found an indication of the complete system with place value.

We have thus seen that the following types of numerals prevailed in ancient India which by evolution gave rise to the modern numerals:

	१	२	३	४	५	६	७	८	९	०
	1	2	3	4	5	6	7	8	9	0
Kharosthi—	I	II	III	IIII						
Asoka—	I	II	III	IIII	+	6				

Nanaghat— 

1 2 3 4 6 7 9

"Cave" 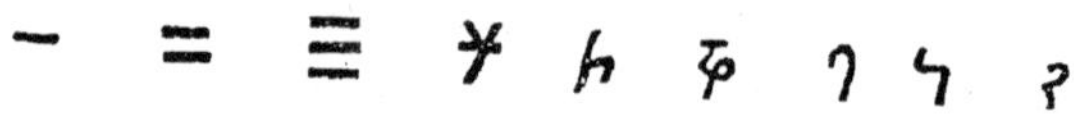

(Nasik)— 1 2 3 4 5 6 7 8 9

Letter— ए द त च प ष स अ न (very primitive).

The last ones are first letters of the words, eka; dvi, tri, catur, panca, sat, sapta, asta, nava.

1 The modern Sanskrit '१' has no doubt originated from, combined with "Asoka" 1.

2 २ from Nanaghat = . One spur mark inside made it two.

3 ३ from Nanaghat ≡. Two spur marks inside made it three.

4 ४ from "Cave" 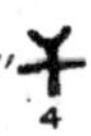 The super-imposition of the last side stroke of three made it four.

5 ५ from प with a spur at the foot, ५.

6 ६ from Asoka "six" with a spur at the foot.

7 ७ from Nanaghat "seven" turned through a right angle.

8 ट from "Cave" "eight" with the tail shortened.

9 ξ from Nanaghat "nine" with the upper part turned, ε from "Cave" "nine" turned.

o ० from circle or from Nanaghat ∝ (ten).

This last theory of the development of the modern numerals is entirely my suggestion. A minute examination of the formation of all the types of ancient Hindu numerals led me to believe in this theory, in order to explain the modern numerical forms.[52]

In this chapter I have tried to trace the gradual development of the Hindu numerals on the distinct and direct testimony of ancient writers on arithmetic.

IV

Among the Arabs

With the accession of the Abbasides, long before the Muhammadan influence spread through Arabia and Persia, passion for learning, especially astronomy and mathematics, grew up with the Arabs. 'The mathematical knowledge of the Arabs rested upon the Elements of Euclid, which, immediately, on being translated into Arabic, was enthusiastically studied by them, and on which they made considerable advances. In the 9th century A.D., they borrowed from the Indians their decimal system, numerals and arithmetic. At the instance of the Caliph Manun (about 280 A.H.) the mathematician Muhammad ibn Musa, commonly called Khowarizimi, wrote a short Algebraical treatise which gave the best known and most useful illustrations drawn from everyday problems of life. This treatise first introduced into. Europe a knowledge of Algebra. The Latin translation of Khowarizimi served as a manual to the European scholars of the 16th century supplying to them knowledge of Algebra earliest in point of time.[53]

From that time onward, there were numerous Arabic writers upon arithmetic, as that subject occupied one of the high places among the sciences, but most of them did not feel it necessary to refer to the origin of the symbols, the knowledge of which might well have been taken for granted.

The earliest Arabic documents containing the numerals are two manuscripts of 874 and 888 A.D.[54] They appear a century later in a work written at Shiraz in 970 A.D. There is also an early trace of their use on a pillar recently discovered in a church apparently destroyed as early as the tenth century, not far from the Jeremiah Monastry in Egypt. A graffito in Arabic on this pillar has the date 349 A.H. which corresponds to 961 A.D.[55] For the dating of Latin documents the Arabic forms were used as early as the thirteenth century. In fact, it was from the eighth century that progress of learning began with the Arabs; from the eighth century to the thirteenth century Muhammadism was to the western world what Rome and Athens and the Italo-Hellenic influence generally had been to the ancient civilisation. Mr. Libri, in his Histoire des Mathematiques, says, if they did not possess the spirit of invention which distinguished the Hindus and the Greeks, they at least possessed the virility of a new and victorious people, with a desire to understand what others had accomplished, and a taste which led them with equal ardour to the study of algebra and of poetry, of philosophy and of language.[56]

Mr. Fleet says that the method of calculation by abacus spread from India to Arabia. It is to be noted that the debt the Arabs are said to owe to Hindu mathematicians for their arithmetical notation is closely connected with the question of the existence of the abacus in ancient India. The gobar numerals of the Arabs are said to be derived more or less from the abacus. Mr. G. R. Kaye[57] has pointed out that while various writers have said that the abacus was in common use in India, they have not given any proof of their assertion. He accepts no doubt that a form of abacus is now in use in India but that fact, he says, does not enable us to affirm that the applicance dates from any early times in India. What is wanted is distinct evidence and Mr. Fleet[58] draws our attention to a passage which perhaps yields such evidence. The passage is found in the Divyavadana which is regarded as dating generally from before A.D. 100, though parts of it may be placed two centuries or more later. It is in chapter 19, entitled Jyotiskavadana (ed. Cowell and Neil, p. 262ff); the story begins thus: —

"There lived in Rajagrha a rich householder named Subhadra, who greatly favoured the Nirgranthas (the Jains). On a certain occasion Buddha wandering through Rajagrha for alms, came to Subhadra's house, and in the course of conversation told Subhadra that his wife was about to bear a son who would devote himself to the Buddhist faith and attain the condition of an Arhat. The interview was witnessed by a person named Bhurika who was the owner of a charitable hall, and was entitled to expect that Buddha should have applied to him for alms rather than to Subhadra. When Buddha, having his alms-bowl filled by Subhadra, had gone away, Bhurika went to Subhadra, and asked what had occurred. Subhadra explainded. Then we are told:— स भूरिको गणित्रे कृतावी श्वेतवर्णां गृहीत्रा गणयितुमारब्धः (He, Bhurika, who was skilled in *ganitra,* took a *svetavarna* and began to count or calculate.).

The upshot was that Bhurika confirmed the prediction. It is clear that Bhurika was an adept in making and testing predictions by means of calculations. The editors, regarding him as an astrologer, have explained "ganitra" in their index of words as meaning "an astrologer's instrument, an abacus," and ''svetavarna" as meaning also "an astrologer's instrument." The appropriate meaning of svetavarna lit. "having a white colour" is "chalk," as assigned to it in Monier Williams' Sanskrit Dictionary. Now what is the meaning of ganitra?. Ganitra is formed with the suffix *itra.* The rule 3. 2. 184 of Panini teaches us the use of this suffix in the sense of instrument to form such words as khanitra, 'an instrument for digging,' aritra, 'a rudder,[59] lavitra, 'a sickle,' and dhavitra 'a fan.' This gives an appropriate meaning to ganitra which explains the above passage thus:

"He, Bhurika, being skilled in the use of the appliance for counting (of abacus) took a piece of chalk and began to count or calculate."

This distinctly suggests the use of some form of the abacus in the shape of a board—the well-known *phalaka*—coloured black and ruled ready for use so that calculations could be made on it with a piece of chalk. This gives, according to Mr. Fleet, a documentary evidence of the use of abacus in ancient India. This was surely the father of the system of gobar or dust numerals. The significance of the term gobar is doubtless that the numerals were written on the dust abacus. It has been stated by Al-Biruni that the Hindus often performed numerical computations in the sand. The term is found as early as circa 950, in the verses of an anonymous writer of Kairwan, in which the author speaks of one of his works on gobar calculation. From the tenth century the Arabs used these gobar numerals and they knew the numerals as Indian forms. It is, therefore, certain that the Arabs used the gobar forms before the later numerals reached them in 773 A.D. The gobar numerals varied more or less, but substantially they were of the following forms (read from right to left):[60]

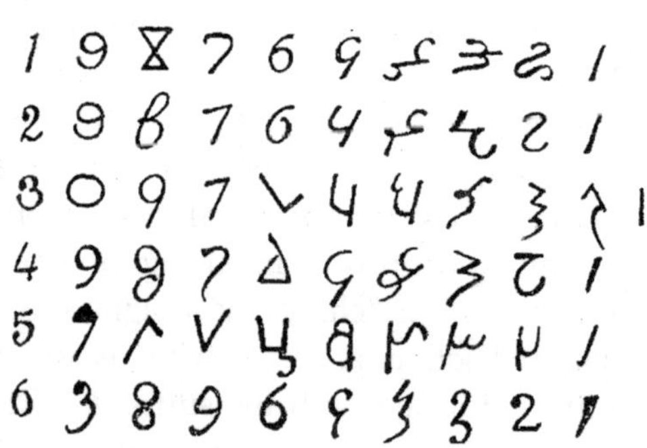

The first definite trace that we have of the introduction of the system of numerals into Arabia dates from 773 A.D. At that time the second Abbaside Khalif Al-Mansur was ruling and an Indian astronomer visited his court bringing with him tables of the planets according to the mean motions, with observations relative to both Solar and Lunar Eclipses and the ascension of the signs; these were translated into Arabic at the Khalif's command by Muhammad Bin Ibrahim Al-Fazari.[61] Al-Khowarazmi and Habash (Ahmed ibn Abdallah, died c. 870) based their well-known tables upon the work of Al-Fazari. It may be asserted as highly probable that the numerals came at the same time as the tables. They were certainly known a few decades later, and before 825 A.D., about which time the original of the *Algoritmi de numero Indorum* was written, as that work makes no pretence of being the first work to treat of the numerals. The three writers mentioned above cover the period from the end of the eight to the end of the ninth century and the historians Al-Masudi and Al-Biruni followed quite closely. One document cited by Woepcke is of special interest since it shows at an early period, 970 A.D., the use of the ordinary Arabic forms alongside the gobar. The title of the work is *Interesting and Beautiful Problems on Numbers* copied by Ahmed ibn Muhammad (951-1024) from a work by a priest and physician, Nazir ibn Yumn al-Quas (died c. 990).[62] It was Al-Khowarizmi who appreciating at once the importance of the position system brought from India wrote an arithmetic based upon these numerals, and this was translated in the time of Adelhard of Bath (c. 1130). This system used by Al-Khowarizmi found its way into Europe through the Latin translation.

In this way when the importance of the place value was appreciated by the Arabs, the complete system of modern Arabic numerals grew up:

١ ٢ ٣ ٤ ٥ ٦ ٧ ٨ ٩ ٠

The Arabs, however, did not adopt the circle, since it bore some resemblance to the letter "five" of their system. The earlist Arabic zero known is the dot, used in a manuscript of 873 A.D. Sometime both the dot and the circle are used in the same work having the same meaning, which is the case in an Arabic Ms., an abridged arithmetic of Jamshid, 982 A.H. (1575 A.D.). As given in this work the numerals are:

I have already stated that the Hindus called the zero Sunya or void. This passed over into the Arabic as assifr or sifr. When Leonardo of Pisa wrote upon the Hindu numerals he spoke of this character as zephirum. Maximus Planudes called it tziphra The English cipher or zero, French chiffre, is derived from the same Arabic wood assifr.

I have already cited the various forms of gobar numerals, an examination of which will show that the 5th form resembles to a great extent the modern Arabic numerals, except that there is no sign for zero in the gobar system. The sign for zero was added much later when the Arabs came to appreciate the importance of place value from the Indian mathematicians. They finally adopted the modern numerals and helped to introduce them into Europe.

V

The Introduction and Spread of the Numerals in Europe Indicating a Solution of the Hindu Arabic Question

We have dealt with the origin and growth of Hindu numerals and also the use of the numerals among the Arabs; here we shall take up the question of their introduction into Europe, a solution of which, we are led to believe, helps the solution of the Hindu-Arabic question of the numerals.

"One theory is that the Hindu system of numerals, without the zero, early reached Alexandria (450 A.D.), and that the Neo-Pythagorean love for the mysterious and specially for the oriental led to its, use as something bizarre and cabalistic, that it was then passed along the Mediterranean, reaching Davius Boethius in Athens or in Rome, and the schools of Spain, being discovered in Africa and Spain by the Arabs even before they themselves knew the improved system with the place value."[63] Another theory has been set forth by Bubnov that the numerals found their way in Europe from ancient symbols used on the abacus, but that the zero was of Hindu origin.[64] The latter theory is not tenable in the light of what has been said in previous chapters.

The Spanish forms of the numerals were called the huruf al-gobar, the gobar or dust numerals, as distinguished from the huruf al-jumal or alphabetic numerals. The signification of the term gobar in doubtless that these numerals were written on the dust abacus, this plan being distinct from the counter method of representing numbers. Al-Biruni states that the Hindus often perfomed numerical computation in the sand.

The gobar numerals themselves were first made known to Europe by Silvestre de Sacy, who discovered them in an Arabic manuscript from the library of the ancient abbey of St. -Germain-des-Pres. The system has nine characters, but no zero. A dot above a character indicates tens, two dots hundreds, and so on, 5 meaning 50 and ½ meaning 5,000. When we consider that the dot is found for zero in the Bakhsali manuscript[65] and that it was used in subscript form, in the Kitab al-Fihrist[66] in the tenth century, and as late as the sixteenth century, although in this case probably under Arabic influence, we are forced to believe that this form may also have been of Hindu origin. The fact seems to be that the Arabs did not immediately adopt the Hindu zero, because it resembled their 5; they used the superscript dot as serving their purposes fairly well; they may, indeed, have carried this to the west and have added it to the gobar forms already there, just as they transmitted it to the Persians. Furthers more, the Arab and Hebrew scholars of Northern Africa in tenth century knew these numerals as Indian forms, for a commentary on the Safar Yesirah by Abu Sahl ibn Tamin (probably composed at Kairwan, c. 950) speaks of "the Indian arithmetic known under the name of gobar or dust calculation".[67] The Indian use of subscript do to indicate the tens, hundreds, thousands etc. is established by a passage in the Kitab al-Fihrist (987 A.D.) in which the writer discusses the written language of the people of India.

The gobar numerals varied more or less, but were substantially, as those shown on page 359 above.

These gobar numerals resemble modern numerals much more closely than the Arab numerals do.

How did these gobar numerals reach Boethius and Europe? Certainly along trade routes with the merchant people trading in India. Even in very remote times, before the Hindu numerals were sculptured in the cave of Nana Ghat, there were trade relations between Arabia and Indus. Indeed, long before the Aryans went to India the great Turanian race had spread its civilisation from the Mediterranean to the Indus. At a much later period the Arabs were the intermediaries between Egypt on the west, and the farther Orient. The Romans also exchanged products with the East. Horace says, "a busy trader, you hasten to the farthest Indies, flying from pover over sea, over crags, over fires."

There may be another theory of the spread of numerals in Europe. With the progress of Buddhism, severals centres of learning set up in India which attracted scholars from China, Persia, Central Asia Greece and other far off countries. In 62 A.D., the then Chinese Emperor sent an ambassador to India.

So that it was entirely possible for the numerals of the Punjab to have worked their way to western countries through Afganistan and Persia at that early date.

As regards the Arabs, in the fifth century they traded by sea with India and even with China, and Hira was an emporium for the wares of the East, so that any numeral system of any part of the world could hardly have been isolated. Long before the activity of the Arabs, ships sailed from the Isthmus of Suez to all the commercial ports of Southern Europe and up into the Black Sea. Hindus were found among the merchants who frequented the bazars of Alexandria, and Brahmins were reported even in Byzantium. The Gulistan of the Persian poet Sadi contains such a passage:—

"I met a merchant who owned one hundred and forty camels, fifty slaves and porters. He answered to me: 'I want to carry sulphur of Persia to China, which in that country, as I hear, bears a high price; and thence to take Chinese ware to Roum; and from Roum to load up with brocades for Hind; and so to trade Indian steel to Halib. From Halib I will convey its glass to Yeman, and carry the painted cloths of Yeman back to Persia.'[68] Still another certain route for the entrance of the numerals into Christian Europe was through the pillaging and trading carried on by the Arabs on the northern shores of the Mediterranean. As early as 652 A.D. in the thirteenth year of the Hejira, the Muhammadans descended upon the shores of Sicily and took much spoil. It is therefore reasonable to conclude that in the Middle Ages as in the time of Boethius, it was a simple matter for any enquiring scholar to become acquainted with such numerals of the Orient as merchants may have used for warehouse or price marks.[69]

To sum up, therefore, there was abundant intercourse between the East and the West for some centuries before the Hindu numerals appear in any manuscripts in Christian

Europe. The numerals must of necessity have been known to many traders in a country like Italy at least as early as the ninth century, and probably even earlier, but there was no reason for preserving them in treatises. Therefore when a man like Gerbert (afterwards Pope Sylvester II from 999 to 1003) made them known to the scholarly circles, he was merely describing what had been familiar in a small way to many people in a different walk of life.

In the 12th century A.D., four Englishmen: Adelhard of Bath (c. 1130), Robert of Chester (c. 1143) William Shelley, and Daniel Morley (1180) are known[70] to have journeyed to Spain in the twelfth century for the purpose of studying mathematics and Arabic. Adelhard of Bath translated from Arabic into Latin Al-Khowarazimi's astronomical tables, while Robert of Chester is known as the translator of Al-Khowarazimi's algebra. The earliest trace of computation with Hindu numerals in Germany is an Algerismus of 1143, now in the Hofbibliothek in Vienna. It was about the same time that Rabbi Abraham ibn Meir ibn Ezra of Toledo wrote a book called the Safar ha-Mispar, the Book of Number, in the Hebrew language. Probably he acquired his knowledge of the Hindu arithmetic in his native town of Toledo.

Of all the mediaeval writers, probably the one most instrumental in introducing the numerals to the scholars of Europe was Leonardo of Pisa. But more popular were the treatises of Alexander de Villa Dei (c. 1240 A.D.) and John of Halifax (Sacrobosco, c. 1250 A.D) who were responsible for the rapid spread of the numerals among the common people. Sacrobosco's work was of special interest in asmuch as the extended use of this work made the term 'Arabic numerals' commonly known. In the introduction it is stated that this science of reckoning was due to a philosopher named Algus, whence the name algorismus, and in the section on numeration reference is made to the Arabs as the inventors of this science. Though some of the commentators were aware of the Hindu origin of the numerals, they took the text as it stood; and hence the belief that Arabs were responsible for the invention of the numerals gained much ground in Europe.

In Europe the complete system with the zero was derived from the Arabs in the 12th century, and the arithmetic based on this system was known by the name of Algorithmus. This word is nothing more than a transcription of Al-Khowarizmi as was conjectured by Reinaud, and has become plain since the publication of a unique Cambridge manuscript containing a Latin translation, perhaps by Adelhard of Bath, of the lost arithmetical treatise of the Arabian mathematician. The arithmetical methods of Al-Khowarizmi were simplified by later Eastern writers, and these simpler methods were introduced to Europe by Leonardo of Pisa in the west and Maximus Planudes in the east. The term zero appears to come from the Arabic sifr through the form zephyre used by Leonardo.[71]

As a result of this brief survey of the evidence mentioned in this Chapter, we come to these conclusions:—(1) commerce and travel helped the spread of numerals in Europe: (2) merchants trading in India, China, Arabia, Persia and Palestine had every opportunity of knowing the elementary number systems of the peoples with whom they were trading, but they did not put this knowledge in a permanent written form; (3) there is every reason to believe that the numerals in some forms were known to European merchants, long before the Arabs introduced the perfect system into Europe; (4) the wonder is that though the Hindu-Arabic numerals were known in 1000 A.D. and formed the subject matter of an elaborate work by Fibonacci in 1202, more extended manuscript evidence of their appearance before that time has not been found, even when they were more or less known to many merchants and travellers of Christian Europe in the Middle Ages.[72]

VI

Hindu Origin of Numerals and the Gradual Evolution of the Modern Forms

It has been said that the earliest Arabic documents containing the numerals are two manuscripts of 874 and 888 A.D. They appear about a century later in a work written at Shiraz in 970 A.D. There is also an early trace of their use on a pillar recently discovered in a church apparently destroyed as early as the tenth century, not far from the Jeremiah Monastery in Egypt. A graffito in Arabic on this pillar has the date 349 A.H., corresponding to 961 A.D. The complete system with the zero was introduced, however, in Europe by the Arabs in the 12th century, and the arithmetic based on this system was known by the name of Algorithmus, algorithm. But the use of numerals in India can be followed back to the Nanaghat inscription, supposed to date from the early part of the 3rd century B.C. The forms of the later Indian numerals for the nine digits appear to be clearly derived from the earlier system. The Nanaghat and "cave" numerals give forms earlier than the introduction of the system of position, while Devanagari, derived from a computation of all the systems, was used with a zero and position value.[73]

"It was not apparently till 773 A.D." says C. E. Bayley, "that the Arabs became acquainted with the Indian ciphers and with the Indian method of notation and arithmetic. They obtained this knowledge from a book presented by the envoy of an Indian monarch to the Khalif Al-Mansur. Probably the Indian monarch was one of the Hindu kings of Kabul. At least the modern Arabic numerals seem to be derived from the peculiar form of those then employed in that part of India. At that date the complete Indian system (employing the zero) was used, it was with this that the Arabs first came in contact. This seems clear from the excessive eulogiums lavished by them upon the new system of numeration and calculation, as being infinitely superior to the Greek system."[74]

As a matter of fact there is abundant reason for believing that Hindu numerals would naturally have been known to the Arabs and even along every trade route to the remote west, long before the zero entered to make their place-value possible, and that the characters, the methods of calculating, the improvements that took place from time to time, the zero when it appeared, and the customs as to solving business problems, would all have been made known from generation along these same trade routes from the orient to the occident. It must always be kept in mind that it was to the tradesman and the wandering scholar that the spread of such learning was due, rather than to the schoolman. Indeed Avicenna[75] (980-1087 A.D.) in a short biography of himself relates that when his people were living at Bokhara his father sent him to the house of a grocer to learn the Hindu art of reckoning, in which this grocer (oil dealer, possibly) was expert.[76] Even in very remote times, before the Hindu numerals were sculptured in the cave of Nanaghat, there were trade relations between Arabia and India. Indeed long before the Aryans went to India, the great Turanian race had spread its civilisation from the Mediterranean to the Indus.[77] As a much later period the Arabs were the intermediaries between Egypt and Syria on the west, and the further orient. Indeed, there is ample testimony to the caravan trade from India, across Arabia, and on to the banks of the Nile. About the sixth century B.C. Scylax, a Persian Admiral under Darius, from Caryanda on the coast of Asia Minor, travelled to north-west India and wrote upon his ventures. Such number systems as there were in these lands would naturally have been known to a man of his attainments.[78]

For over five hundred years Arabic writers and others continued to apply to works on Arithmetic the name "Indian."

In the tenth century such writers are 'Abdallah-ibn Al-Hasan, Abul Qasim (died 987 A.D.) of Antioch and Mohammed ibn Abdallah Abu Naser (c. 982) of Kalwada near Bagdad. Others of the same period or earlier (since they are mentioned in the Fihrist, 987 A.D.) who explicitly use the word "Hindu" or Indian, are Sinan ibn Al-Fath of Harran and Ahmed ibn Omar Al-Karabisi. In the eleventh century came Al-Biruni (973-1048) and Ali ibn Ahmed Abul-Hasan Al Naswi (c. 1030). Even in the first half of the 14th century, Maximus Planudes, a Greek monk, followed the Arabic usage in calling his work "Indian" Arithmetic.

The next question is, therefore, the source of information of the Arabs. In the seventh century the Arab empire became an ellipse of learning with its foci at Bagdad and Cordova, and its rulers not infreqently took pride in demanding intellectual rather than commercial treasures. It was at this time that the Hindu numerals found their way to the north. It is not possible to find out the exact time when they were actually known. But this much is certain that in the eighth century they were taken to Bagdad.

It is stated in the preface to the Astronomical tables of Bin-al-Adami, published by his continuator, Al Qasim in 920 A.D., that in the reign of the second Abasside Khalif Al-Mansur, in 773 A.D. "an Indian astronomer, well versed in the science he professed, visited the court of the Khalif, bringing with him tables of the planets according to the mean motions, with observations relative to both Solar and Lunar Eclipses, and ascensions of the signs. The Khalif embracing the opportunity thus presented to him, commanded the book to be published for a guide to the Arabians." The task devolved on Mohammed Bin Ibrahim AlFazari, whose version is known by the name of the greater Sind-hind. Colebrooke was of opinion that the Sind-hind was a copy of the revised Brahma-sidhanata of Brahmagupta, and that the fact was deducible from the number of elapsed days between the beginning of the planetary motions and the commencement of the present age of the world according to the Indian reckoning as it is quoted by Abu Mashar (an astrologer of Balkh) and which agrees precisely with Brahmagupta.[79] The work of AlFazari, taken from the Hindu astronomy, continued to be in general use among the Arabs until the time of Al-Mamun for whom it was epitomised by Mohamed Bin Musa Al Khowarizmi; and his abridgment was thence forward known by the title of the less Sindhind. Contemporary with Al-Khowarizmi and working under Al-Mamun was a Jewish astronomer, Abul-Taiyib Sened ibn Ali, who wrote a work on Hindu arithmetic. Abul-Hasan Ali Al-Masudi (d. 946) of Bagdad who travelled to the China Sea on the east, at least as far south as Zanzibar, and to the Atlantic on the west, speaks of the nine figures with which the Hindus reckoned.

To sum up, therefore, we have up till now shown that the Indian claim to invention of the value of position and the zero rests first on the distinct testimony of Arab historians, and other Arab writers to that effect; on the certainty that it was practically used by the Indians at a date considerably anterior to that at which it can be really shown to have been used by any other people; and the Indian claim to the "zero" rests on exactly similar grounds. This claim is further establised by the early records of inscriptions on caves and hills during the reign of Asoka and other princes.

As to the forms of the numerals, fashion played a leading part until printing was invented. This tended to fix these forms, although in writing there is still a great variation.

1. In the early printed books "one" was often written as i, perhaps to save types, just as some modern typewriting use the same character for 1 and 1. One appears in such forms as

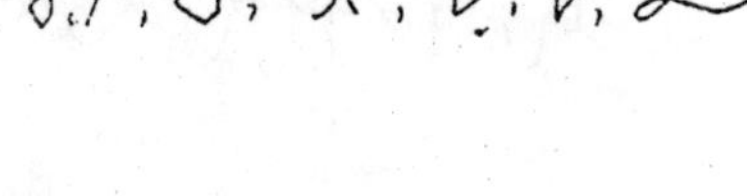

2. "Two" often as z. In early printed books 12 appears as 1 z. In the mediaeval manuscripts the following forms are common:—

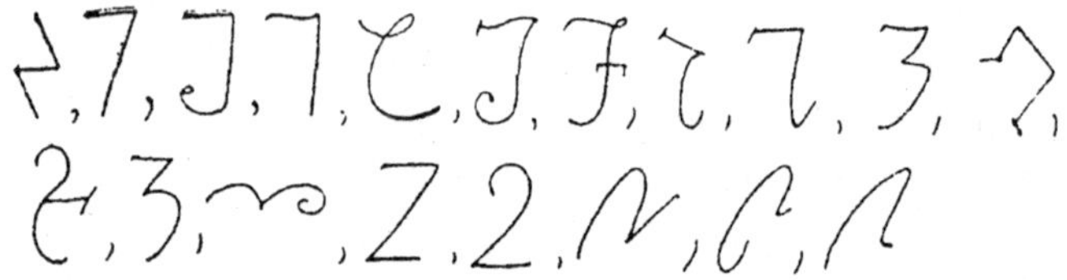

It is evident from the early traces that it is merely a cursive form for the primitive =, just as 3 comes for ≡, as in Nanaghat,

3 "Three" usually had a special type in the first printed books, athough occasionally it appears as 3. In the mediaeval manuscripts it varied rather less than most of the others:

4. "Four" has changed greatly; and one of the first tests as to the age of a manuscript on arithmetic, and the place where it was written, is the examination of this numeral. Until the time of printing the most common form was 9, although the Florentine manuscript of Leonardo of Pisa's work has the form 4.; but the manuscripts show that the Florentine arithmeticians and astronomers rather early began to straiten the first of these form up to forms like 9, and 9 or 9 more closely resembling the present form. The first printed books generally used the present form with the closed top 4, the open top used in writing 4 being purely modern. The following are other forms of the four, from various manuscripts:—

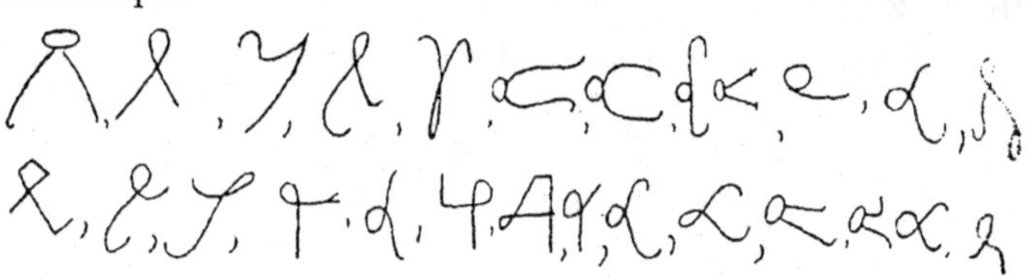

5 "Five" also varied greatly before the time of printing, as:—

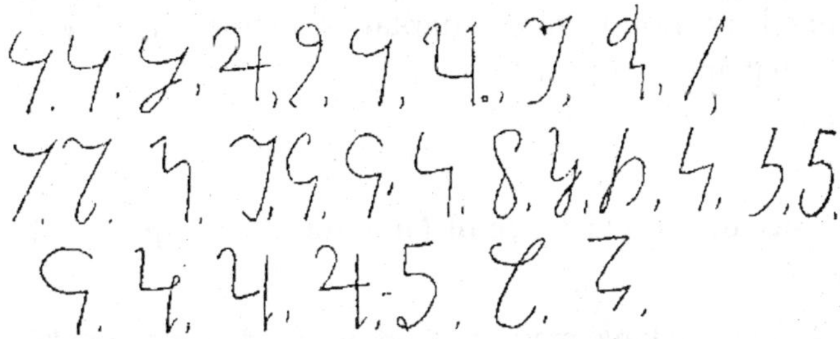

6 "Six" has changed rather less, than most of the others, chief variation being in the slope of the top

7 "Seven" like "four" has assumed its present form only since the fifteenth century.

8 "Eight" like "six" has changed but little. In mediaeval times there are a few variants of interest.

9 Nine has not varied much.

o Variation of zero is given below.

The following are the variations.

7 

8 

9 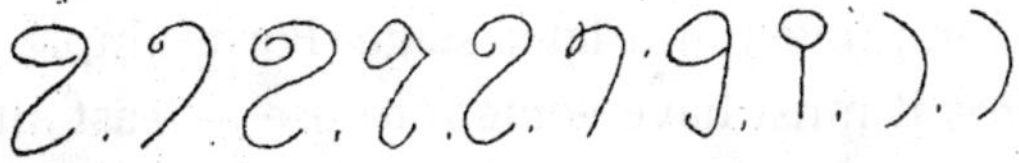

0 

Thus we see that the present forms 1, 2, 3, 4, 5, 6, 7, 8, 9 and o have been obtained through a series of variations, obtained, no doubt, from adaptation of the Indian types of Nanaghat and Cave numerals.

VII

Evidence of the Existence of Numerals with Space Value in Literature

So long we have mainly based our evidence on the "Cave" and "'Hill" inscriptions and have proved that numerals with space value existed among the Hindus as early as the sixth century. Dr. D. E. Smith in his "History of Mathematics" (vol. II) and in "Hindu Arabic Numerals" by himself and Mr. L. C. Karpinski, has also given the credit to the Hindus for the existence of numerals with space value among them at such an early date as the sixth century. But here we shall show that there is ample proof in Sanskrit and Pali literature regarding the existence of numerals even at an earlier date. Though figure numerals can only be found in inscriptions, yet there is sufficient evidence in literature to warrant the existence of a perfect system of numerals at a date much earlier than that of the inscriptions.

We have already seen that a system of numeration existed at the Vedic period i.e. about 2000 B.C. In the Upanisads also there is mentiọn of *aksarapuja* (worship of letters) which indicates the existence of letters and words at the time of the Upanisads (1400 B.C.). From this time probably the word and letter system of numerals began to be used, though it was extended and improved at a later stage. For example, Rama stands for *three;* it must have come into use at least after the birth of Balarama (the third Rama) i.e. after the Puranic age. But *aga* (mountain) standing for *seven, abdhi* (sea) standing for *four,* and *agni* for *three* must have been used before the Pauranic age; as we get in the Puranas अष्ट कुलाचलाः (eight mountains and not seven) and सप्त समुद्राः (seven seas and not four).

This word-and-letter system of numerals was used even in the sixth century A.D. by Varahamihira in his Pancasiddhantika and Brhatsamhita; and this system was the most popular one generally used by the masses. Then why is it that Aryabhata did not use the word-and-letter system and used alphabetic numerals as क, का, कि, की, कु, कू, कृ, कृ etc.? Because, we think Aryabhata, a great scholar himself, wanted to introduce a simpler system to show his learning. In spite of this attempt to innovate a new system he could not but use the word *kha* (void, sky) to mean zero, as in the following,—

वर्गाक्षराणि वर्गेऽ वर्गेऽ वर्गाक्षराणि कात् ङ् मो यः।
खद्विनवके स्वरां नव वर्गेऽ वर्गे नवान्त्यवर्गे वा।।

Thus the Hindus had three sets of numerals:

1 Word-and-letter system. 2 Alphabet system. 3 Figure system.

In Sanskrit literature also there is ample evidence of a perfect system of numerals even before the sixth century A.D. In Sankara's commentary on the Brahma-sutras there is a passage: —

Yatha caikapi sati rekha, sthananyatvena nivesyamanaikadasasatasaha sradisabdapratyayabhedam anubhavati (Sariraka Bhasya, 2, 2, 17) i.e. as the same stroke (rekha from lekha meaning line or stroke) coming in different positions gives the idea of units, tens, hundreds, thousands etc. This clearly shows the existence of the knowledge of place value at the time of Sankara. Now what is the approximate date of composition of Sankara's commentary? Orthodox Hindus believe from the inscriptions in the archives in the Mathas (monasteries) that Sankara lived in the first century B.C.; K. T. Telang fixes the date in the sixth century A.D.;

Pathak followed by Wilson fixed the date at 778 A.D. Tilak found a mistake in the calculation on astronomical basis of Pathak and fixed a date a century earlier i.e., at 678 A.D. The popular view is that Sankara lived in the sixth century A.D. and died at the age of forty. Any how the system with place value must have been very popular in the sixth century, otherwise Sankara would not have illustrated his point by a reference to the system of place value. Moreover there is a passage in the Vyasa Bhasya of the Yogasutras of Patanjali to imply that "the same stroke is termed one in the unit place, ten in the ten's place, hundred in the hundred's place." This clearly refers to the place value system. The date of Vyasa Bhasya has been fixed in the early part of the sixth century by all oriental scholars. Therefore, it is certain that the system with place value was known before the sixth century probably one or two centuries earlier, so that within that time the system became so well-known that it served the purpose of reference even in philosophical dissertations.

There is also some reliable evidence of the existence of numerals with place value even before the Christian era. In Kautilya's Arathasastra, book I, chap. V, it is said regarding the education of a prince that "after the ceremony of tonsure, which is generally performed at the third year, (cf. Raghu, iii, 28) he is taught lipi and samkhyana.[80] Samkhyanam here surely means numerals. Also in book II, chap. VII, where the business of keeping accounts in the office of the Accountant is mentioned, it is stated that the districts were divided into different *chaukies* and there was a perfect system of sending weekly, fortnightly, quarterly, half-yearly and annually accounts to the head-quarters. "Accounts will be submitted in the month of Asadha. When they (the accountants of different districts) present themselves with sealed books, commodities and net revenue, they shall all be kept apart in one place so that they cannot carry on conversation with each other. Having heard from them the totals of receipts, expenditure, and net revenues, the net amount shall be received. Checking of accounts will be made by the minister in charge."

This examination of accounts pre-supposes a perfect system of numeration. For it could not have been by means of sticks as in Europe even till the 15th century, records of which are found in old monasteries. It is absurd to think that so many sticks could be gathered and checked at one place. There must, therefore, have been a perfect system of numeration at that time. Now the age of Kautilya is fourth century B.C. So that it is not unwarrantable to conclude that a system of numerals existed even in the fourth century B.C.

The question then arises why did not Asoka coming after two generations (as his grand-father, Candragupta, was a contemporary of Kautilya) use this system? Probably because it was not very popular even then; and as Asoka wanted publicity of his inscriptions he resorted to the popular system, and also wrote his inscriptions in Prakrit. Further he did not lay much stress on dates.

In the poetical portion of Lalitavistara (x, 15) also it is stated in the tenth chapter that "lipi (alphabet) and sankhya (numerals) are to be learnt by the prince. The poetical portion of *Lalitavistara* dates from the first century B.C.

This, however, does not settle as a matter of course the date of invention of the zero as numeral. Strictly speaking these illustrations do not mention whether the place value was indicated in writing numbers or was being indicated on the abacus. In other words it may be asked whether it will be safe to conclude that the zero was known in that early age. Hence instances of the use of the zero from any early writing will be at once decisive and settle the question altogether. We have an instance in Subandhu's Vasavadatta (C. 620 A.D.)[81]

and one in Sriharsa's Naisadhiyacarita (C. 12th century)[82] where the zero has been expressly stated to be sunyabindu meaning a point or a dot. Better instances are furnished by Varahamihira's Pancasiddhantika (505 A.D.) where he incidentally states two fundamental arithmetical operations by the zero (kha, sunya, ambara) viz addition and subtraction in more than one place, e.g., chap, iii, 2,' 17; chap, iv, 7, 8, 11, 12; chap, xviii. 35, 44, 45, 48, 51. It is note worthy that all the above verses are from those chapters of Panca-siddhantika where are summarised the teachings of the Paulisasiddhanta Thibaut remarks: "Varahamihira has in no case obliterated the characteristic features of the siddhantas he has to deal with, and whatever distinguishes those works from one another in the text of the Pancasiddhantika really distinguished them in their original form."[83] Unfortunately no means of verification of this conjecture has been left to us in the present age, the book not being now extant. It was known to Brahmagupta (c. 628 A.D.), to Bhattotpala (966 A.D.), and even as late as in the eleventh century to Al-Biruni. There are, however, numerous quotations in Bhattotpala's excellent commentary on Brihat Samhita[84] from an "original Paulisa-siddhanta" and probably also from a different edition of the same work. In any case this conjecture will lead us to the conclusion that the zero was known in India before 400 A.D., for that is the higher limit for the time of composition of the original Paulisasiddhanta as set by Thibaut.[85] It may further be stated that Whitney believed that in the time of Aryabhata (476 A.D.), the Hindus had "invented their system of signs employed in decimal notation."[86]

The earliest instance of the use of zero is found in Chandah-sutra of Pingala or Pingalanaga, which is a manual of Vedic metres. One section of the book deals with the problem of determination of the number of possible variations for a metre with a given number of syllables. It is in this connection that Pingala has used the word sunya in two successive sutras (chap. VIII, S.29-30). These two sutras, in fact the whole of the manual, are found re-stated in the Agni-Purana (chaps. 328-34). There is no doubt that, by sunya, Pingala was referring, not to the mere concept of nothingness, but to a definite symbol whose concept is akin to the concept of our zero numeral. Halayudha, the earliest commentator of Chandahsutra, has explained the sutras fully and has adduced illustrative examples as well. He is of opinion that, in the sutras referred to, sunya denotes the zero, and he has been supported by Weber who remarks that there can be no doubt about that. Now Pingala lived about the middle of the second century B.C., and his book was popular among the Brahmanas, the Vedic scholars of the age. This will be at once evident from its being raised to the level of a Vedanga from its being incorporated in the Agni-Purana and also from its being quoted in the Bharatiya Natya-sastra (chap. XV) and in the Parisistas (Max Muller, Ancient Sanskrit Literature, p. 75). Hence it has to be admitted that the zero was known to the Brahmanas of India in the second century B.C. The date of invention is probably a century or two earlier. There are also in the Atharva Veda, xix, 22, 23, two hymns where there is reference to the zero as well as to positive and negative numbers. In those hymns the zero has been called ksudra (trifling)[87] and positive and negative numbers have been denoted by rca and anrca respectively.[88]

From these instances from Sanskrit literature it is not unsafe to conclude that the numerals with zero existed even before the Christian era, at least in the first century B.C. in a perfect form. No records of any other country can furnish any instance of the existence of numerals at such an early date.

References

1. Even now used in the course of uttering Gayatri mantras to remember the number of times.
2. The system is described by Nicholas Rabda of Smyrna. Vide Palmer in the Journal of Philology, vol. II, pp. 247 ff.
3. Maximus Planudes (c. 1330) states that "the nine symbols came from the Indians."(Waschke's German translation, Halle 1878, 3).
4. From the 1558 edition of the Ground of Artes, fol. C, 5.
5. G.R. Kaye, "Notes on Indian Mathematics—Arithmetical Notation," Journal and Proceedings of the Asiatic Society of Bengal, vol. Ill, 1907, p. 489.
6. Al-Biruni's 'India,' vol. I, chap. XVI.
7. The Hindu name for the symbols of the decimal place system.
8. Sachau's English edition of the Chronology, p. 64.
9. Huart's History of Arabic Literature, Eng. Edition, New York, 1903.
10. W. Brennand's Hindu Astronomy.
11. Varahamihira's Brhat-Samhita, edition of Dr. H. Kern, p. 50.
12. Journal of the Royal Asiatic Society, Vol. XIV, p. 14.
13. Burnell's South Indian Palaeography.
14. Indian Antiquary, vol. XX, p. 54.
15. Taylor's translation of Lilavati, p. 11.
16. In spite of any documentary evidence in his favour, Mr. Kaye has, in season and out of season, so much harped upon this view in his writings on Indian Mathematics that it has to some extent prejudiced the impartial thinking of even scholars like Dr. D.E. Smith.
17. Sir E.C. Bayler, Genealogy of Numerals, Journal of the Royal Asiatic Society, vol. XIV, p. 335.
18. D.E. Smith and L.C. Karpinski—Hindu Arabic Numerals, p. 22.
19. Pandit Bhagavanlal Indraji—"On Ancient Nagari Numeration: from an Inscripton at Nana Ghat." Journal of the Royal Asiatic Society, Bombay Branch, vol. XII.

20. Burnell, South Indian Paleography, 1874, and E. Senart, "The Inscriptions in the caves at Nasik," Epigraphia Indica, vol. VIII, pp. 59-96.
21. C.E. Woodruff, "The evolution of modern numerals," Amer. Math. Monthly, 1909.
22. Vide Prof. S.K. Ganguly, Alp System of Aryabhata, Bull. Cal. Math. Soceity, vol XVII, no. 4.
23. Sudhakara Dvivedi, Ganakatarngini (1892) p. 5: Kaye, *JASB.*, (1907), p. 478; Fleet, *JRAS.*, (1911), p. 125.
24. *JRAS.*, (1911) p. 125.
25. Macdonell and Keith, Vedic Index, vol. I, pp. 342, 353.
26. Heath, History of Greek Mathematics, vol. I, p. 234.
27. Yajurveda, chapter XVII, mantra 2.
28. Kaye, *JRAS.*, (1910), p. 752.
29. *Ibid.*, p. 754.
30. This fact has so far escaped the notice of distinguished oriental scholars owing to their attention having been too much attracted by Aryabhata's alphabetic system. Accordingly Mr. Kaye has been inconsistent in translating Aryabhata's verse giving the modern notation. Fleet perceived this inconsistency and tried to be consistent. But in doing so he has mistranslated the principle *"sthanat sthanam dasagunam syat."*

    Mr. Kaye writes that Jamblichus (4th century) "had perfectly clear ideas on the value of position" and gives the following example in his support:

    "If the digits of any three be added together, and digits of their sum be added together, and so on, the final sum is six." *(IASB.*, p. 493).
31. Mr Kaye attributes this example to Jamblichus and cites as reference Gow's History of Greek Mathematics. But a comparison of Mr. Kaye's statement of this example with the statements of Gow (p. 98) and Heath (History of Greek Mathematics, vol. I, pp. 114 and 115) will show (i) that Mr. Kaye is not justified in putting the example in the above form which seems to support his conclusion and (ii) that

other peoples who regarded numbers as being made up of a certain number of units, a certain number of tens, a certain number of hundreds, etc., had equally "clear" (?) ideas on the value of position

32. *JRAS.*, (1911) page 125.
33. *Ibid.*, page 118.
34. These systems do not include the so-called 'word-symbol notation which is not notation at all but which gives numbers expressed in the modem notation by stating the digits (one or two at a time) beginning with the unit's place.
35. According to Aryabhata's scheme there are no letter numerals for 1 and 2 to occupy a-varga places. It will be seen from what follows that for the figures 3, 4, 5, etc., up to 10 there are two sets of letter-numerals—one set for varga places and the other for a-varga places.
36. Vide Sir Richard Temple's article on the Burmese system of arithmetic, Indian Antiquary, vol. XX, pp. 53-69.
37. Prof. S. K. Ganguly, "Alphabetic System of Aryabhata," Bull. Math. Society, vol. XVII, no. 4.
38. Aryabhata's rules for the volumes of a tetrahedron and a sphere. (vide *JRAS.*, 1911, p. 118).
39. For this interpretation I am indebted to Dr. Bibhutibhusan Datta, D. Sc of the Science College of the Calcutta University.
40. Buhler, Paleographie, p. 82.
41. Burnell, South Indian Palaeography, p. 79.
42. Lilavati, chap. I, verses 2, 3.
43. D.E. Smith and L.C. Karpinski, Hindu Arabic Numerals.
44. Pancasidhantika, chap. I, verses 8, 9, 10.
45. Buhler, Palaeographie, p. 78.
46. "The Gurjaras of Rajputana and Kanauj", —Journal of the Royal Asiatic Society, January and April, 1909.
47. Vol. ix, 1908, p. 248.
48. Thibaut, p. 71.
49. D.E. Smith and L.C. Karpinski, Hindu Arabic Numerals.

50. From a letter to D.E. Smith from G.F. Hill of the British Museum.
51. D.E. Smith in the Bibliotheca Mathematica, vol. IX (3), pp. 106-110.
52. Also Burnell, South Indian Palaeography, 1874.
53. Literary and Scientific Activities under the Caliphate by S. Khoda Buksh.
54. Karpinski, Hindu Numerals in the Fihrist, vol. XI, pp. 121 ff.
55. Archaeological Report of the Egypt Exploration Fund for 1908-1909, London, 1910, p. 18.
56. Libri, Historie des Mathematiques, vol. I, p. 174.
57. J.A.S.B., 1908, pp. 293-297.
58. J.R.A.S., 1911, pp. 519-21.
59. See Buhler, Indian Paleography, pp. 5, sec. 37, c.
60. Silvestre de Sacy gives the ordinary modern Arabic forms, calling them *Indian*.
61. Colebrooke, Essays, vol, II, p. 504.
62. D.E. Smith and L.C. Karpinski, Hindu Arabic Numerals.
63. D.E. Smith and L.C. Karpinski, The Hindu Arabic Numerals.
64. Bubnov, Origin and History of our Numerals (Russian), 1968.
65. Possibly as early as the third century A.D.
66. Ascribed by the Arabic writer to India.
67. Reinaud, Memoire sur l' Inde, p. 399.
68. Gulistan-Sadi, born at Shiraz in 1193. He himself travelled from India to Europe.
69. D.E. Smith and L.C. Karpinski, Hindu Arabic Numerals.
70. Wallis, Algebra, 1685, p. 12 seq.
71. Encyclopaedia Britannica, 11th edition, Numerals.
72. D.E. Smith and L.C. Karpinski, Hindu-Arabic Numerals.
73. Burnell, South Indian Paleography, 1874.
74. Sir C.E. Bayler; "On the Genealogy of Modern Numerals. Journal of the Royal Asiatic Society, vols. XIV, XV.
75. Carrade Vaux, Avicenna, Paris 1900. His full name was Abd-Ali-al-Hossein ibn Sina.

76. D.E. Smith and L.C. Karpinski, Hindu-Arabic Numerals.
77. G. Oppert, On the Ancient Commerce of India, Madras, 1879, p. 8.
78. D.E. Smith and L.C. Karpinski, Hindu-Arabic Numerals.
79. Colebrooke, Essays, vol. II, p. 505.
80. Vrttacaulakarma lipim sankhyanam copayunjita.
81. Hall's edition, p. 181.
82. Canto I, stanza 21.
83. Introduction to Pancasiddhantika, p. xvii; cf. p. xvi.
84. Bhattotpala's commentary, edited by Dvivedi, Benares (1895).
85. Mr. G.R. Kaye says that the Paulisasiddhanta is an adaptation of a work by Paulus Alexandrinus, a Greek astronomer. But the two works are entirely different in as much as the latter work deals with astrology while the former with astronomy. Vide an article by the present author on Hindu Astronomy in "East and West," Aug. 1919.
86. Journal of Amer. Oriental Soc., vol. 6, p. 563 footnote.
87. Cf. Amarasimha's synonym for zero as 'tuccha', meaning insignificant, negligible. Amarsimha is a lexicographer of the 5th century A.D.
88. For the concluding portion of this chapter, I am indebted to an article by Dr. Bibhuti Bhusan Datta of the University College of Science, Calcutta, which was published in the American Mathematical Monthly, Nov., 1926.

Sukumar Ranjan Das

5
The Present Mode of Expressing Numbers

In our present mode of expressing numbers with the help of nine numerals and the Zero, the notational places of successive higher denomination, such as tens, hundreds, thousands, etc. are arranged from right to left. This fact led Mr. G. R. Kaye[1] to conjecture that it has been imposed upon the modern world by a people with right to left script. And this consideration, amongst others, has led him to suggest that the credit of invention of the modern numerals cannot be due to the Hindus whose scripts from the earliest times are written from left to right. Some weakness in the above hypothesis has been admitted by Kaye himself. For certain scripts are known to have changed their directions. But apart from that, the whole hypothesis is based upon wrong conceptions and unsound and insufficient observations of the modes of expressing numbers by the various nations of the world. It betrays, on the whole a lack of breadth of view and, in particular, ignorance of the method of writing out numbers in full which is found in the Sanskrit and kindred literatures.

In our present study we shall treat separately of the two principal methods of expressing numbers, viz. (1) by writing out the number names in full, and (2) by means of symbols or signs. We shall further confine our attention to the number systems only of the prominent civilised peoples of the world, ancient or modern, such as the Egyptian's Babylonians, Greeks, Romans, Hebrews, Syrians, Persians, Arabians, Turks, Hindus, Chinese, Japanese and Tibetans, etc.

In writing out numbers in full almost all the nations adopted the decimal scale and the whole vocabulary of the numeral language of any nation was very small. It consisted of separate names of nine numerals and of certain denominations, such as tens, hundreds, thousands, etc. It is noteworthy in this connection that while most of the ancient nations did not go beyond the fourth, or at most a fifth denomination, the ancient Hindus even in the remotest Vedic Age (before 3000 B.C.) dealt freely with no less than eighteen denomination.[2] In modern times also, the numeral language of no other nation is as scientific and has attained as high a state of perfection as that of the ancient Hindus. Again while the numeral vocabulary of the Hindus has remained rich and practically full from the earliest times, that of other nations has swelled by addition of a very few more names only. So in the latter there has been a cumbrous system of grouping and regrouping for coining the necessary terminologies. A number expression is formed with the help of these names mainly by addition and partly by multiplication. The subtractive principle also is in evidence in certain specific instances of Sanskrit number names; for example, nineteen is expressed either as 10 + 9 or as 20-1; twenty-nine as 20 + 9 or 30-1; and so on. In later times this has become the usual principle in those cases.

In an additive system it is immaterial, how the elements of different denominations, of which a number expression is composed, are written, for the value of a symbol is quite independent of its position. But it has become the usual custom from olden days to adhere to a definite mode of arrangement, instead of writing in a haphazard manner. The arrangement is this: when a number is composed of the first two denominations only, that is, for names of numbers *below* hundred, the smaller element is written first. This is true of Sanskrit, Zend, Greek, Latin, Arabic, Persian, Turkish, Chinese and some other languages. But when a number expression contains higher denominations as well, these are written in *descending* order in front of the number names already formed. For instance if a number is composed of first four denominations, the normal mode of writing it will be first thousands, then hundreds, then units and then tens. Thus there is a sudden change of order in the process of formation of the number expression. We shall not attempt to search for any probable explanation of this change of order. It will be, in fact, of no use to the object in view. We shall simply take the facts as they are found, analyse them, and discover the normal process of formation, if any. It is very striking that this change of order in the process of formation of number expression is common to most of the important languages. Only in a very few languages, such a English,[3] Tibetan, etc., the order is continuously descending. There is, however, one language in which it is sometimes permissible to write all the denominations in the *ascending* order. This speciality belongs to the Arabic. And it is over and above the normal process already indicated and as an alternative to it. Save this, it thus appears to be the commonly accepted normal process for the expression of larger numbers in the languages of the various nations of the world to write the elements of higher denominations before the lower.[4] The Sanskrit and kindred languages contain certain other minor irregularities. But those are exceptional and rare cases. They were noted and sufficiently explained by the celebrated grammarian Panini (c. 700 B.C.). In any case, it would not be proper to cite them to show that the normal process of formation of number expressions in Sanskrit is different from that indicated above. There can be cited from the Sanskrit literature, innumerable instances in which the larger elements are placed first systematically. Thus truly speaking, it is only a lack of sufficient observations that led Kaye to assert that in Sanskrit the smaller elements are usually placed first.

The scripts of different races are written in different directions the Semitic races write from right to left; the Aryans from left to right and the Mongolians from the top downwards. It has already been stated that certain scripts are known to have changed directions. But there has never been any serious deviation from the normal process of number expressions as indicated above. It has remained universal. And this is very remarkable indeed.

We shall now pass on to the problem of numeral notation or symbolism. Though a discussion of the forms of the symbols, their origin variation and mutual relation will be highly interesting and instructive by itself, it will be beyond the scope of the subject matter of the present study. So we shall avoid it as far as possible and simply confine ourselves to the critical examination of the law of arrangements and other incidental affairs. Theoretically speaking, the symbols should be as few and simple as possible, if not fewer and simpler they should not exceed the number names. That is, nine symbols for the first nine number-names together with some symbols for the denominations ought to have sufficed, by the application of the same multiplicative and additive principles as in number terminologies, to express all the numbers of a language. But in actual practice number symbolism was not so simple. The difficulties would be many times more in the case of those languages in which there were no separate simple names for all the denominations. We have seen that, except in the Sanskrit and kindred languages of India, those defects are present in all other languages, for denominations above thousand. Even for numbers with lesser denominations, the difficulties appeared no less insurmountable. For almost all the ancient peoples failed to invent even nine separate symbols for first nine numerals. For smaller numbers, it looked as if ingeneous and advantageous. More symbols mean, indeed, heavier tax on memory. But in case of larger numbers, it required repetition to an unwieldy extent. Repetition could be minimised to a certain extent by an early recognition of the multiplicative principle on a little wider scale. While majority of the ancient nations recognised it for the multiples of hundred, they failed to do so for multiples of ten. Their mind was more bent upon inventing separate symbols for all the tens. To minimise repetition, recourse was taken to change of the scale of notation and other devices. Thus while the scale of notation in the numeral languages is found to have remained decimal all throughout, that in number symbolism is found in some cases to have become sexagesimal as well. In this way, number symbolism became highly complicated. Estrangement of plans between numeral names and numeral notations was visible from the earliest stage. In fact, before the advent of the Hindu place value system, the problem of number expressions by signs was many times more complex and different than the problem of number expression by writing out in full. But in one matter they proceeded on the same principle; in expressing numbers in terms of two or more symbols of different values or denominations, the larger element was put first. And this was followed consistently and uniformly all throughout. So much so that the change of order which was, and still is, visible in the normal process of formation of number expression by names completely disappeared from the normal process of number symbolism.

The theoretical plan of numeral notation indicated above is found in actual practice in the Sinhalese and the modern Chinese number symbolism. The nearest approach to it was made in the Alphabetic numeral system of the Hindu Aryabhata (the elder, born 476 A.D.), in which the notational places were indicated by the vowels, and the pure consonants signified the numerals.[5] It had, however other defects and digressions and hence was not adopted by people.

Of the numeral symbols of the ancient Semitic peoples, the Egyptian Hieroglyphic had separate signs for one, ten, hundred, thousand, ten thousand, hundred thousand and million; the Phoenician for one, ten, twenty and hundred; the Palmyrene for one, five, ten and twenty; the Syriac for one, two, five, ten, twenty and hundred; the Hieratic and the Demotic for one, five to ten, and twenty, thirty up to hundred. The symbols for two hundreds, three hundreds, etc. were formed, except in the case of the Hieroglyphic, by putting the smaller element in front to denote multiplication. The numbers were expressed by repetition and addition of these symbols as necessary. As was characteristic of the Semitic scripts, the numbers were written from right to left; and it began with the largest element. The Hieroglyphic numbers were also written left to right and sometimes, as in the early inscriptions, from top downwards. It is interesting to note that in the former case the pictorial numeral signs were turned the opposite way.

The ancient Babylonians had relatively smaller number of numeral signs, viz., for one, ten and hundred. The last one was really a combination of the other two. In fact they seem to have only one wedge-shaped symbol, which signified two numeral values in two different positions. There were, however, also instances of separate symbols for one and ten. Thus the Babylonian numeral system was the poorest of all the numeral systems of the ancient civilised nations. It was also the crudest. For owing to the paucity of separate symbols, a comparatively greater amount of repetition was unavoidable. And above all, which was its worst feature, the numerical value of a symbol was not fixed. It stood either for a minimum value (one or ten) or for 60 multiple of the same. In each case, the numerical value of a symbol had to be determined from the context. The brightest aspects of the Babylonian number symbolism were the invention of the sexagesimal scale of notation and the introduction of a symbol to indicate the vacant denomination in it. In writing numbers, on either scales, the Babylonians put the larger elements before the smaller.

The Roman numbers were expressed in terms of symbols for one, five, ten, fifty, hundred, five hundreds, and thousand. The normal mode of writing was to put the greater elements in the beginning. In certain cases it was reversed to denote subtraction. But on a longer view, that is, in a compound number of still greater value, it will be found that the normal mode has been preserved on the whole.

The Greeks had two systems of numeral notation—the Attic system and the Alphabetic system. The former was no better than the systems that have been discussed so far. Only a slight innovation was in evidence in the symbol for fifty which was devised on a multiplicative principle from the symbols of five and ten. No other numeral system, probably except the Palmyrene, contained this principle in symbols for numbers smaller than hundred. The Attic system was replaced about 400 B. C. by the other system. The use of the letters of alphabets for the purpose of numerals was common with the Hebrews, the Syrians, the Arabs, the Persians and other Semitic peoples. The Arabs continued the practice even for many centuries after they became aware of the Hindu numerical system.

It was probably the Greeks who first conceived the happy idea of making a systematic use of the letters of the alphabet as numerals.[6] Some investigators are of opinion that the idea originated with a Semitic people and the Greeks only borrowed it. It is known that the Greeks obtained their alphabets from the Phoenicians. But that latter did never use their alphabets with numerical significance. One thing is, however, true. The Greek alphabets (24 in number) were insufficient for a satisfactory system of alphabetic numerals. So they appropriated three Phoenician alphabets. It has been stated by Al-Biruni that the Arabs got their system from the Hebrews; so did probably the Persians and others. In this system the letters of an alphabet are divided into three groups; the first group of nine letters were used to signify nine units; the second group of nine letters to denote the nine tens; and the third group were used to signify the hundreds. The numbers were expressed on either scale, the decimal as well as the sexagesimal. In writing a compound number the larger element, or the element of the higher denomination was put first. This was as much true of the peoples like the Greeks who wrote from left to right, as of the Semitic peoples who wrote from right to left. But in certain Greek inscriptions of Asia Minor the smaller elements were put first and in a few others the arrangement was irregular. These facts were probably not without any significance. They will lend additional support to those who believe in the ultimate Semitic origin of the alphabetic system of numerals. For the number expressions written by a Semitic people in their normal descending order are apt to be looked upon by the Greeks as arranged in the ascending order. The primitive Greeks not only borrowed the system but also copied the arrangement which appeared at the first sight. The irregular arrangement probably referred to a state of transition. But ultimately the normal Greek mind prevailed and the mode of the number expressions was changed specially to bring them to conformity with the mode of the other Greek system, viz. the Attic System.

Attention has already been drawn to the one great peculiarity of the Arabic numeral language, that in writing out a number in full, it was permissible to follow either the ascending or the descending order. But in their notations they consistently followed the descending order. There are numerous applications of alphabetic numerals in Al-Biruni's *Chronology of the Ancient Nations.*[7] In every instance the higher denomination has been put before the lower. These facts deserve more than a passing notice. About the middle of the 8th century A.D., there appeared amongst the Arabs a very ingenious system of writing numbers with the help of nine signs and the Zero. The Arabs called them *al-arqam al-hind* ("the signs of the *Hind*"). The Syrians obtained the system a century or a little more earlier. From there, the system was introduced into Europe near about the 12th century. And this is now the commonly accepted numeral notation of the modern world. The numbers according to this system are written by the Arabs from left to right contrary to their mode of writing the scripts, or to the mode of writing their numbers whether in *abjad* or in sexagesimal notation. This is certainly contrary to the general nature of the Arabs and other Semitic races. And this fact alone leads to the strong presumption that this system of numerals came to the Arabs and other Semitic peoples, from a non-Semitic people.[8]

The Chinese have three systems of numerals for use in different connections. Of these two are indigenous; they are written from the top downwards and the elements are arranged in the descending order. The third and the present system has been introduced by the Catholic missionaries. It has been stated before that it agrees with theoretical number system suggested. It is written from left to right with the higher elements before the lower. The numeral systems of the Koreans and the Japanese are derived from those of the Chinese. So there is nothing new to be said about them.

We shall now take up the Hindu numeral notations. The two earliest notations were the Kharosthi and the Brahmi. The Kharosthi was particularly the script of North-Western India (including Afganisthan and the Northern Punjabi). It disappeared in the third century A.D. In the Kharosthi numeral system, there were separate signs for one, four, ten, twenty and hundred. They were written from right to left, with the higher elements before the lower. But in cases of the hundred and its multiples, the symbols for the smaller elements were placed before the symbol for the hundred to denote multiplication. All these were common with the principles underlying other systems of Semitic numeral notation which we have discussed before. The next but more important and largely used Hindu numeral notation was the Brahmi. It was probably earliest numeral notation that was found in India proper and remained in the field for many centuries. The Brahmi numeral system consisted of separate signs for all the units leaving two and three, all the tens, some hundreds and some thousands, including ten, twenty, and seventy thousands. These latter mean symbols for numbers of the fifth denomination which the Greeks called the *myriad.* Thus the Brahmi numeral system is the richest of all the ancient independent numeral systems in symbols. The plan was, indeed, nearly the same as those of the Demotic and the Hieratic; but it excelled the either. Though one or two signs of the hundredth and thousandth denomination might be suspected to have been invented on the multiplicative principle, it was not very apparent and had not been applied systematically. In such circumstances it will be futile to conjecture any relation about the origin of the Brahmi numerals with those Egyptian numerals. Many knotty points will have to be cleared before such a hypothesis can be established. The Brahmi numerals, like the alphabets, were written from left to right beginning with the higher elements. There are also instances, though rare, of the very ancient Brahmi characters being written from right to left. But there was never a deviation from the universal mode of putting the higher elements before the lower.

The various systems of number notation which have been discussed above, whether of the Hindus, or of the Semitic peoples, or of others races, did not contain the idea of place value.[9] Such a system requires, as will be apparent, multiplicity of signs. Again there cannot be put a limit to the number of signs in any system. It is bound to increase with the demand for representing larger numbers which again are concomitant of the growth of civilisation. But the inventive genius of the people of any age and clime failed in devising new signs to cope with the demand. So repetition became an unavoidable necessity, however undesirable it might be. With the development of the Society, energy of the elite was directed towards bringing it to a minimum. But that did not prove to be easy and simple. Certain people conceived the idea of getting out of that difficulty by using the signs already invented for alphabets, also for the purpose of numerals. Apart from other drawbacks, the resourcefulness of such a device, is evidently bound to be limited. This was improved to certain extent by putting bars, dashes or other signs with the alphabets. The introduction of the idea of local value extremely simplified the problem number symbolism.

The third and the modern Hindu numeral system contains only ten signs, the nine numerals and the Zero. With the application of local value they are quite sufficient to express any number however large. The scale is, of course, decimal. In this system the notational places increase from right to left. The numbers are written from left to right in accordance with the Hindu scripts and the higher denominations, as is universal, are always put before the lower. This system is identical with the system which, as we have seen, appeared amongst the Syrians and the Arabs during the early centuries of the christian era. From there it went into Northern Africa and Europe. And this system has now been adopted by all the civilised peoples of the world.

The origin of this system of numeral notation is still shrouded in mystery. As all the numerous alphabets of the Hindus, though differing very widely in form, are believed to have been derived from the Brahmi alphabets, it is highly probable that the modern Hindu numerals also grew out of the old Brahmi numerals. Only three new signs were introduced into the old system: two of which helped to make repetition unnecessary; while a third, the most important of all, helped to discard a lot of other signs and thus reduced the number of symbols to the minimum. There is no gainsaying in the fact that the difference of the modern Hindu numeral system from the various other ancient numeral systems is least in the case of the Brahmi. From the mere fact that the notational places increase from right to left, —it has been already stated in the beginning—Kaye conjectures that the modern Hindu numeral system is not really of Hindu origin, but of Semitic origin. Such a conjecture is undoubtedly wrong. Because for a people with left to right script, such an order of the notational places is a natural concomitant of the custom of writing the higher elements before. From what has been stated above, it will be abundantly clear that this custom is universal. Kaye may think the present order "inconvenient and clumsy". But the world is known to be following it from time immemorial, even when the place value system of decimal notation was not invented. For peoples with right to left scripts, the order suggested by Kaye, that is the reverse order, would have been natural. But those peoples, while adopting the modern place value system, did not change the order of the notational places they found amongst the people from whom they borrowed. And to keep up the time honoured custom of writing the higher elements first, they rather began, strangely enough to write numbers according to the new system from left to right, contrary to the nature of their scripts and also to the direction of writing their numbers according to the other numeral systems known to them from before, e.g., the *abjad* and the sexagesimal notation. It is interesting to note in this connection that the same normal mode of writing numbers was prevalent amongst the Mayas of Central America who had a fairly well developed system of numeral notation with the place value. The Mayas followed the scale of 20, except in one step and they had a symbol for the Zero. Again, the present mode of writing numerals follows most closely the mode of expressing numbers by writing out the names in full which is current in the Sanskrit and kindred literatures from the remotest Vedic ages. From the latter, only the names of the denominations have been deleted, for they are sufficiently indicated by the position of the digit in the series. Thus it is established beyond doubt the present system of decimal notation cannot be suspected to be of non-Hindu origin simply on the score of the mode of writing numbers.

We shall conclude the present paper by drawing attention to what the Hindu Scholiasts have said on the subject under discussion. The works of most of them are still in mess. and I have had access to a very few. The subject attracted the attention of some of the eminent scholiasts of the celebrated Hindu mathematician Bhaskara. The earliest of them, as far as is known to me, to notice the subject was Ganesa *(c.* 1545 A.D.).[10] It was also noticed by Krisna *(c,* 1575 A.D.), the court astronomer to the Emperor Jehangir of Delhi, and by his nephew Munisvara (born 1603 A.D.) and by Narsimha (born 1568 A.D.).[11] All of them are of the same opinion. As, of the successive denominations tens, hundreds, thousands, ayutas, etc. each one is more respectable *(abhyarhita)* than the preceding ones, it should be written first, for it is natural to give first place to the more respected one. For a people with left to right script, that will be possible if the denominational places are arranged to increase from right to left.

References

1. G.R. Kaye, "Notes on Indian Mathematics—Arithmetical Notation," *J.A.S.B.,* vol. iii (1907), pp. 475-508; *Indian Mathematics,* Calcutta (1915), p. 32.
2. *Vide* Macdonell and Keith, *Vedic Index,* vol. I, pp. 343 f.
3. Excepting the expressions containing elements smaller than twenty.

4. Cf. Karl Fink, *Geschichte der Elementar-Mathematik,* authorized English translation by W.W. Beman and D.E. Smith, *Brief History of Mathematics,* Chicago, 1910, p. 8; also Hankel *Geschichte der Mathematik,* p. 32.
5. Most of the modern interpreters of the first Aryabhata's alphabetic numeral notation have failed to grasp its true significance. This has been exposed correctly by Pandit Durgadas Lahiri (*Prthivir Itihasa,* Calcutta, vol. iii, pp. 332 et seq) and Mr. Saradakanta Ganguly [*Bull, Cal. Math. Soc.,* vol. xvii (1927), p. 202] and it is quite in accordance with the interpretation of his commentators.
6. In India the letters of the alphabet were used to denote numbers as early as the seventh century B.C. But it seems that it never passed beyond a preliminary crude stage for several centuries. Anything of the kind of a system appeared, as far as is known, only about the fifth century A.D. and even then it was not much in vogue. It is also significant that while the systems of alphabetic numerals of all other peoples were similar, those of the Indian were on different plans. Compare A. Weber, *History of Indian Literature,* English translation by Mann & Zachariae, London (1878), p. 222.
7. *Vide* Al-Biruni's *Athar-ul-Bakeya* (or "Vestiges of the Past) This book has been published in the Arabic original (Edward Sachau, *Chronologie Orientalischer Volker von Alberuni,* Leipzig, 1878) and in the English translation (C. Edward Sachau, *The Chronology of the Ancient Nations,* London, 1879).
8. Cf. John Taylor. *Lilavati,* Bombay (1816), Introduction, p. 31 sq.
9. The principle of local value was more or less in evidence in the Babylonian sexagesimal notation and in the notation of the Mayas of Central America.
10. *Vide* Ganesa's commentary on *Lilavati,* ii. 12.
11. Vide *Siddhanta Siromani,* edited by Muralidhara Jha with Munisvara's *Marici* and Nrsimha's *Vasanavartika,* Benares, 1917, pp. 57, 89.

Bibhutibhusan Datta

6

Cosmographical Theories of the Hindu Astronomers

From the earliest stage of the development of human knowledge attempt has been made to fathom the mysteries of creation and to link by a common principle of metaphysical and scientific investigation the different members of the solar system. Now in proceeding with this enterprise they encountered two distinct problems. One was concerned with the nature of the primeval World-stuff; the other with the operation to which it had been subjected. Modem theorists have made it their primary object to expound the mechanism of cosmic growth, the play of forces involved in it, the transformations and progressive redistributions of energy attending it. But early thinkers till the time of Descartes tried to solve this question by assuming an appropriate material for the exercise of their constructive ingenuity.

Thales asserted all things to have been derived from water. Anaximenes substituted air. A crude attempt to determine the cosmic origin is also found in the Rg-veda, where it is stated that at first darkness prevailed everywhere, it was only chaos and water abounded in the whole creation.[1] The Manu Samhita also asserts that the creator at first created water and then fire and life.[2] The Vortex Theory of Descartes or the Modern Nebular Theory of Kant and Laplace ultimately reduces itself to the theory that the universe originated from vapour as nebulae are no better than vapour.

Hindu astronomers have begun the discussion on cosmographical theories with a series of questions regarding the earth, its magnitude and divisions, the situation of the seven Patala Bhumis or imaginary lower regions of the earth, the sun's revolutions the causes of day and night of the Gods, the Demons and the Pitrs, the order of the stars and planets, the position of their orbits with respect to each other in the Universe.[3] Next they speak of imperceptible agencies of creation, almost the same metaphysical theories as may be found in the Vedas, the Puranas and other mythological works.

It is said, "At first only darkness prevailed; in that darkness Vasudeva (in whom the whole Universe lay latent), the Supreme Being whose manifestation is everything that exists, who is transcendental, without attributes, and tranquil, who is beyond the twenty-five primordial matters and inexhaustible, who pervades all places within and without,[4] who is the contracting power (lit., he who contracts) first created water at the beginning and threw his own power (of contraction) in it. That water with the power of contraction in it produced a golden egg. The sides of the egg were even then enveloped in darkness. Out of that egg evolved Aniruddha (whose motion cannot be stopped) Sanatana (who is everlasting). Hence he is called Hiranyagarbha in the Vedas; as he was the first to evolve out he was called Aditya and for the creation of the Universe he is called Surya (the Sun). Surya (the Sun), whose another name is Savita, who dispels darkness and who is the cause of the creation, existence and destruction of all beings, is always moving and brings to light the different worlds....... From this originated the moon, the five planets, stars, the earth and other worlds. Brahman, the creator, exists within that egg and hence that egg is called Brahmanda (the Universe). In the hollow of that egg is situated this world consisting of Bhu, Bhuvas etc. and not outside that egg. It is round (गोलकाकृतिः), as if two cauldrans are joined together with faces in contact.[5]

Though the above theories seem to be more or less metaphysical, yet they can be given a scientific interpretation. It is seen that अप् was first created or was at first in existence. अप् is generally translated as water but it also means watery vapour or simply vapour. The word अप् also indicates ether and hence it means ethereal vapour. Hence it can easily be understood to the same as nebulae. Thus as first it was only darkness and ethereal vapour or nebulae all around. Into that was thrust the force of contraction and the result was the outcome of a golden (सुवर्ण) egg. Utpalabhatta meant by this word "luminous".[6] In the Manu Samhita in connection with the theory of creation of the Universe Kullukabhatta, the commentator, has clearly said that it was not golden but brilliant like gold (शुद्धिगुणयोगवशात्). The visible universe is shaped like an egg, it is not exactly round. By the force of contraction, stars, planets, the sun and other orbs are formed. The force acts within the egg in a state of constant rotation. That is to say, the ethereal vapour which existed at first produced this universe by the forces of contraction and rotation.

The explanation given above is not far-fetched or imaginary. It is the most natural explanation. Then what is the difference between the above theory stated in the Surya Siddhanta and the Nebular Hypothesis of Kant and Laplace?

The Brahmanda or the golden egg of Brahma is the vast hollow sphere of the Universe at the centre of which is the earth; within it all the stars are supposed to revolve daily, and beneath them are the orbits of the planets Saturn, Jupiter, Mars, Venus, Mercury, the Sun and the Moon, in the order of the distances from the centre.[7] The earth stands firm at the centre by its own power without other support in space.[8]

In this connection the Hindu Astronomers dwelt on some peculiar geographical theories which have now probably lost their earlier mythological significance and seem to be pure figments of the imagination.

The circumference of the sphere of the Brahmanda, to which the solar rays extend, is declared to be equal to the product of the moon's revolutions (57,753,336000) in a Kalpa by the circumference of the moon's orbit. Bhaskara says, "Some astronomers assert that the length of the circumference of the sphere of the universe is 18712069200000000 yojanas. But those by whom the astronomical science was thoroughly understood say that it is the length of the circumference of the sphere up to the limit the darkness-dispelling rays of the sun extend. In my opinion every planet covers so much distance in yojanas in one kalpa. Hence it is called Vyomakaksha (व्योमकक्षा)".

Varahamihira has given further details of the constitution of the Universe in his Pancha Siddhantika:

"The round ball of the earth, composed of the five elements, abides in space in the midst of the starry sphere, like a piece of iron suspended between magnets; covered on all sides with trees, mountains, towns, groves, rivers, oceans and other things, in its middle there is Sumeru the abode of the gods and below there is the place of the Asuras, Straight above Meru in space one pole is seen; the other pole is seen below, placed in space. Fastened to the pole the sphere of the stars is driven round by the pravaha wind.

Beneath the equinoctial circle is Lanka; there the sphere is right Day and night there are always of the same length, viz. 30 nadikas.

Of the moon which is constantly placed below the Sun, one half is illuminated by the sun's rays, while the other half is obscured by the moon's own shadow, as is the case with a jar standing in the sunlight. The rays of the sun, being reflected from the moon which consists of water, destroy the darkness of the night, just as the rays of the sun falling on the surface of a mirror destroy the darkness inside a house.

Above the moon there are Mercury, Venus, the Sun, Mars, Jupiter and Saturn (in succession), and then the stars. All planets move towards the east with the same velocity, each in its own orbit. The planets arranged in the ascending order upwards from the moon are the Lords of the months (in succession); in their descending; order downwards from Saturn, they are the Lords of the hours; if we take each fifth member of the ascending series we have the Lords of the days. The Lords of the year have been explained before."[9]

We have thus given a complete account of the cosmographical theories and the constitution of the universe as stated by the Hindu astronomers. However crude might these theories have been, they were surely the forerunners of the Vortex Theory of Descartes and the Nebular Theory of Kant and Laplace.

Sukumar Ranjan Das

Kumarila and Dinnaga

Any one who is acquainted with the Indian philosophical texts will admit that the controversy between the Buddhists and the Hindu philosophers appears most prominent in the history of Indian culture. But it is a matter for regret that nothing has yet been done to disentangle the various threads in the web of this controversy. The neglect of this field is not, however, without reason. The authors of the texts either make a passing reference to the views of others or quote them without informing us either of their source or of their author. The commentaries, too, sadly lack the information we seek for. They mention only occasionally names of philosophers who hold the opposite views without, however, referring to the work or works from which they have drawn their quotations. Owing to ravages of time many valuable books that were current at the time among literary circles have been lost, and the meaning of many passages of the texts we read today have therefore become obscure. We do not know the proper setting of the views that are found controverted in these books. This is perhaps one of the reasons why the full significance of the controversies has still remained a mystery to us. Fortunately for us, some of the texts that have been lost in Sanskrit have been preserved in Tibetan and Chinese translations. The only thing that we could do under the circumstances is to make a comparative study of the Hindu and Buddhist texts that exist in Sanskrit, Tibetan, and Chinese.

As an illustration we may now refer to the *Slokavartika* of Kumarila and its Tika, the *Nyayaratnakara* of Parthasarathimisra, and compare some of their passages with those of the *Pramanasamuccaya* of Dinnaga now available only in Tibetan translations. The name of Dinnaga is well-known to the readers of the *Nyayavartika* of Uddyotakara with its Tika by Vacaspatimisra. The pages of the *Nyayaratnakara* (Chaukhamba Skt. Series) abound in questions from the works of a Bhiksu.[10] There are also direct references to Dinnaga.[11] Kumarila himself once refers to him as *Nyayavid*.[12] A comparative study of these passages with those of *Pramanasamuccaya* would convince the reader that Kumarila hurls his vehement attacks here against the views of Dinnaga; and the terms *Bhiksu, Bhavadvraddha,* and *Bhavadiya*[13] occurring in the *Nyayaratnakara* are but impersonal references to the Buddhist philosopher, Dinnaga. This is made still more explicit in the chapter on *Apoha;* for there it seems as if Kumarila had by his side the *Pramanasamuccaya* of Dinnaga when writing that chapter of his Vartika.

The parallel passages are given below without nothing occasional differences:

	Nyayaratnakara	*Pramanasamuccaya*[14]
(1)	सदित्यसद्रव्यु दासाय न नियोगात् स गंस्वते। संप्रयोगो हि नियमात् सत एवोपपद्यते।। 4. 36; p. 144	gal te yod pas med bsal na । sbyor las de ni rtogs par hgyur । yan dag sbyor ba nes par ni । yod nid la ni bstan pa yin ।। 1; $37^{c.d}$ —$38^{a.b}$.
(2)	सान्तरग्रहणं न स्यात् प्राप्तौ स्थानाधिकस्य वा। 4, 41; p. 146.	bar dan bcas pa hdzin pa dan । ses pa lhag pahan thob mi hgyur ।। 1; $20^{c.d}$.
(3)	बुद्धिजन्मनि पुंसश्च विकृतिर्यद्यनित्यता। अथाविकृतिरात्माऽयं प्रमातेति न युज्यते।। 4.52; p 150.	skyes bu rnam par hgyur nas blo । gal te bskyed na mi rtag hgyur ।। ci ste bdag la hgyur med na । de la tshad ma hthad ma yin ।। 1; 48,
(4)	कल्पनाऽपि स्वसंवित्ताविष्टा नार्थे विकल्पनात्।। 4. 134; p. 175.	rtog pahan ran rig nid du hdod । don la ma yin der rtog phyir ।। 1; $7^{a.b}$.
(5)	मैकं रुपाद्यभेदो चेन्नेन्द्रियेण तत्। अक्षाऽनेकत्ववैयर्थ्यं स्वार्थे भिन्नेऽपि शक्तिमत्।। 4. 156; p. 183.	geig min gzugs sogs geig tu gyur । mthon na dban po las di min । dban gz'an don med hgyur phyir ro । ran yul tha dad kyan hdzin nus ।। I; 25^{b}-26^{a}.
(6)	दृष्टान्तोऽभिहितो ह्येष वैधर्म्येणाऽसुशिक्षितैः। 5. 154; p. 254.	mi mthun phyogs la legs bslabs pas । dpe de nid du brjod pa yin । II; $103^{c.d}$.
(7)	परार्थानुमानं तु खट्टष्टार्थ प्रकाशकम्। 5, 146; p. 252-53.	gz'an gyi don gyi rjes dpag ni । ran gis mthon don gsal byed yin । III; $98^{a\text{-}b}$.

(8) ज्ञानान्तरेणाऽनुभवे हीष्टा तत्रापि च स्मृतिः ।	ses pa gz'an gyis nams myon na ।
विषयान्तरसंचारस्तथा न स्यात् स चेष्यते ।।	thug med la han dran pa ste ।
5.27, 187; pp. 277, 321.	yul gz'an dag la hpho ba na । med hgyur de yan hdod phyir ro ।। II; 13.
(9) पूर्ववत् फलमस्य ।	hbras bu snar bz'in hdi gnis kyi ।
5.52; p. 361.	II; 49[e].
(10) तिरुपाल्लिङ्गतोऽर्थदृक् ।	tshul gsum rtags las don
5. 49; p. 488.	mthon paho । II; 49[b].
(11) न जातिशब्दो भेदाना–	rigs sgra khyad par rnams la min ।
मानन्त्यव्यभिचारतः ।	mthah yas phyir dan hkhrul pahi
वाचको योगजात्योर्वा	pyir ।
भेदार्थै रपृथक्श्रुतैः ।।	brjod byed Idan nam rigs tha dad ।
51, 114; p.596.	don dan tha dad med thos phyir ।। V; 171.
(12) तद्दतो नाऽस्वतन्त्रत्वात् ।	ran dban med phyir de Idan min ।
5.120; p.598.	V; 173[a].
(13) तद्वांस्तु भेद एवेष्टः	de Idan kyad par nid la brjod ।
स तु पूर्वं निराकृतः ।	de yan snar ni spans pa yin ।
तद्दन्मावेऽपि सम्बन्धः	de Idan tsam ni hbrel ba ham ।
मत्ता वेति विचारितम् ।।	yod pa yin zes rnam par dbyed ।।
5.128, 131; pp. 600-601.	V; 178.
(14) तद्वानर्थो घटादिश्चे–	de ltar de ni bum sogs bz'in ।
न्न घटादिषु वर्तते ।	bum pa la sogs rnams la spyi ।
सामान्यमर्थः स कथ–	mi hjug de don ji Itar yin ।
मिति तद्वदवाच्यता ।।	de yan rgyu mtshan med mi hdod ।।
5, 131; p.601.	V; 179.

(15) मा भूदर्थान्तरे वृत्ति–		geig la dkar po min sogs bz'in ।
र्वद्येकत्राऽसितादिवत्		V; 180[a].
	5.131; p. 601.	
(16) नैवं जातेरजातितः		de min rigs la rigs med phyir ।
	5. 133; p. 602	V; 180[c].
(17) अर्थाक्षेपेऽप्यनैकान्तः		don gyis hbans kyan ma nes so ।
	5. 133; p. 602.	V; 180[d].
(18) भेदो भेदान्तरार्थं तु		khyad par don gz'an khyad par ni ।
विरोधित्वादपोह्यते ।		hgal bahi phyir na sel bar byed ।।
	5.147, p. 605.	V; 197[a-b].
(19) अन्यत्वेऽपि न सामान्य–		spyi dan khyad par rnam grans sgra ।
भेदपर्यायवाचिनाम् ।	5.149, p. 606.	gz'an nid yin yan sel mi byed ।।
		V; 194[c-d].
(20) अदृष्टत्वाद् व्युदासो वा		yan na ma mthon phyir sel to ।
भेदानामितरेतरम् ।		khyad par gyi ni spyi bsal hgyur ।।
	5. 155; p. 608.	V; 200[a-b].
(21) नाऽन्ययुक्तस्य दृष्टत्वात्		ma yin gzan ldan mthon bahi phyir ।
	5.155; p. 608.	V; 200[c].
(22) तन्मात्राऽऽकाङ्क्षणाद् भेदः		de tsom hdod phyir khyad par ni ।
तत्सामान्येन नोज्झितः ।		ran gi spyi yis spon mi byed ।
	5.155; p. 608.	V; 195[a-b].
(23) वृक्षत्वपार्थिवद्रव्य–		sin nid sa las gyur rdzas yod ।
संज्ञेयाः प्रातिलोम्यतः ।		ses bya go rim bzlog pa las ।
चतुस्त्रिद्व्येकसन्देहे		bz i gsum gnis dan gcig the tshom ।
निमित्तं निश्चयेऽन्यथा ।।		gz an du des la rgyu mtshan yin ।।
	5.158 p. 609.	V; 204.
(24) जातिधर्मव्यवस्थितेः ।		rigs kyi chos ni rnam gnas phyir ।
	5, 163; p. 611.	V; 206[a].

Apastamba and Gautama

Buhler (S.B.E. 2, intro.) declared that Gautama is the oldest Dharmasutra now extant and Jolly in his "Recht und Sitte" has accepted this theory (p. 6). I differ on this point and will try to show that excepting the dubious evidence of the Caranavyuha there is nothing to prove that Gautama is older than Apastamba; all probabilities on the contrary seem to indicate rather just the opposite—that Apastamba is older than Gautama.

First and foremost, the fact must not be lost sight of, that Ap. was pre-Paninian. His work must have been written at a time when the Sanskrit language was not in that state in which it was found by Panini. Secondly that he was not far removed in time from Svetaketu, the celebrated teacher of the Satapatha Brahmana. He might have been an older contemporary even, for Ap. rejects his theory about the study of the Veda after marriage without the slightest show of ceremony (I,4,13,20) and Buhler has proved that this Svetaketu is without doubt the Svetaketu of Sat. Br. (S.B.E. 2, xxxviii). Now the Aitareya Br., even on the most conservative computation, cannot be dated later than 800 B.C. Taking this to be the *terminus a quo* for the date of the Sat. Br. and the age of Katyayana as the *terminus ad quem* (see his Varttika to Panini iv, 3, 105), the Sat. Br. cannot be dated later than 600 B.C. and all things considered, Ap. must be dated about 500 B. B. Buhler too arrived at practically the same result (S.B.E., 2, XL, 111). Ap.'s contiguity to the age of the Brahmanas may also be inferred from another peculiar feature of his Dharmasutra—its very frequent references to the various Brahmanas. This is seen in no other Dharmasutra.

Regarding Baudhayana's priority to Ap. it may safely be said that the alleged references to Baudhayana in Ap., upon which the whole theory is based, are in no way convincing. It requires not a small amount of ingenuity to discover them.

Ap. labours to controvert the authority of a Vedic passage which has been quoted in Baudh. Can it reasonably be concluded from it that Ap. is posterior to Baudh? Ap. mentions by name not a few authorities on Dharma, of which some Dharmasastras in revised version, are still in existence. Why not then assume—if it is at all necessary—that one of these authors had quoted this Vedic passage as well? As for the second alleged down to the field of law we find that the legal concept of Ap. are strangely meagre and puerile. Gaut; (X. 31) solemnly lays down the law of ownership, also found in later Dharmasastras, but nothing of the kind is known to Ap. Gaut. (XII, 29ff) gives various laws about different rates of interest, pledges and deposits, closely resembling those of later Dharmasastras and he has even no objection to a Brahmana lending out money at interest, provided that he does it through an intermediary (X, 6), but Ap. uncompromisingly prescribes punishment for one who "lends money at interest" (1, 9, 27, 10) and declares the food offered by a usurer unacceptable (1, 6, 18, 22). One of the most striking features of Ap. is that the custom of imposing fines for crimes is not known to him. Punishments prescribed by him mostly amount to mere threats of hell and damnation; Dandaniti proper is a sealed book to him. But Gaut. on the other hand prescribes various fines (XII, 8ff.) and gives the correct grammatical derivation of the word *danda* (XI, 28). In conformity with the principles of later Dharmasastras, Gaut. gives, laws as to how long a wife should have to wait for her absent husband (XVIII, 15ff.); Ap. is absolutely reticent on this point. The fact that Ap. depends much more upon custom than any other Dharmasutra is a proof of his early care. His last sutra speaks volumes in favour of his high antiquity, in which he frankly confesses that the remaining duties should be learnt from men and women of all castes. Gaut. (XXI, 7) once refers 10 Manu, Ap. never. Gaut. (XIX. 14) knows various places of pilgrimage but Ap. is quite innocent of them.

Arguments may thus be multiplied, but I think sufficient has already been said to prove the priority of Ap. to Gaut. Now, if the theory of interpolation is carried so far as to cover all these points, it amounts to saying that the Gaut. which had preceded Ap. is no longer extant but there is nothing to show that there actually was any such, and at all events it must be admitted that the Gaut. as we have it is later than Ap.

Lastly I beg leave to point out that the supposed acquaintance of Ap. with the division of Hindu learning as taught in Madhusudana Sarasvati's Prasthanabheda,—taken by Buhler to be and indication of his late origin (SBE, 2, XXIX-XXX), is but a myth. Ap. II, 11, 29, 11 is one of the passages in translating which Buhler has been led astray on account of his excessive reliance on the interpretation of the commentator. This Sutra declares that "the knowledge which Sudras and women possess is the completion (of all study)." The commentator takes this knowledge of Sudras and women to be "the knowledge of dancing, acting, music and other branches of the Arthasastra" and according to Buhler, this interpretation is "without doubt, correct" (SBE, 2, XXIX). Now, Arthasastras, as they are known to us, do not teach dancing and music and it is the unanimous verdict of Grhya and Dharmasutras that members of the upper castes should never devote themselves to these profane arts whether before or after the study of the Veda. Moreover even without any help from outside it may be proved that what Ap. here has in view is acara and not dancing and music. In the sutra immediately following he declares, "this knowledge is a supplement of the Atharvaveda" and in the next says: "It is difficult to learn the sacred law from (the letter of) the Vedas (only); *but by following the indications it is easily accomplished*". Now who can doubt that in the Sutra no. 11 Ap. lays down that on completing the study of the Veda one should learn acara from Sudras and women? Very probably the specific mention of the despised

Sudras and women in this connection thus sadly misled the great savant. But as Ap. (II, 6, 15, 10) expressly declares that rites for the dead should have to be learnt of women and that duties must be learnt from women and *men of all castes*" (II, 11, 29, 15), there can be no doubt that Ap in in passage concerned has nothing but acara in view.

References

1. Rg-veda, X, 129.
2. Manu Samhita, chap. I.
3. Suryasiddhanta, chap. XII, verses 1-9.
4. Cf. Rg-veda (Purusa Hymn), x, 90, 1.
5. Surya Siddahnta chap. XII, 12 to 29 verses.
6. Brhat Samhita—The chapter on Upanayana (उपनयन), comment on verse 6.
7. Surya Siddhanta, Chap. XII, verses 30, 31.
8. Siddhanta Siromani, Goladhyaya, Chap, I, verse 2.
9. Panca Siddhantika, Chapter XIII.
10. (a) तत्र भिक्षणा, p. 144. (b) भिक्षुणा विषयैकत्वतृष्णाया, p.361.

    (c) यत्त भिक्षुणा पक्षधर्मेत्यार्दि, p. 377.
11. (a) दिङनागगवचनमपि न्यायमूलत्वात्, p. 250.

    (b) अत एव दिङ्नाचार्यैरप्युभयं, p. 253.

    (c) दिङ्नागगवक्षनमपि, p. 259.

    (d) दिङ्गागस्त्विदं प्रत्यक्षविराधोदाहरणं, p. 365.
12. वृद्धानामप्यस्मदीयानामिदमेवाऽभिमतमित्याह—न्यायविङ्गिरिति, p. 258.
13. (a) भवद्वद्भैरेव दिङ्नागाचार्यैः, p. 250.

    (b) भवदीयेनापि, 250. (c) तदा भवदीयात्, p. 488.
14. The Xylograph used by me belongs to the Visvabharati Library.
15. The text appears to be corrupt here.

Batakrishna Ghosh

7
Aspects of the History of Hindu Astronomy

"A man", says Seneca, "can hardly lift up his eyes towards the heavens without wonder and veneration to see so many millions of radiant lights and to observe their courses and revolutions, even without any respect to the common God of the Universe." Thus even at the dawn of human consciousness when the light of knowledge was fitfully glimmering from within the veil of ignorance, the science of Astronomy, the most ancient of all sciences, attracted attention. Even from the earliest ages, the beauty and the grandeur of sunrise and sunset and of the dreamy mystery that the night presents, excited awe and wonder even in the most careless observers and naturally a spirit of enquiry was roused in them:

"O Almighty omniscient Lord, what is the measure of this earth? What is its shape? Who holds it firm? What are its divisions? Where are the seven regions on it? How does the cycle of night and day originate from the Sun? And does he revolve illumining the different worlds?" (*Surya Siddhanta*).

The basis of Hindu Astronomy was in the religious beliefs of Hindu votaries in times when each heavenly body represented a Divinity. The study of Astronomy originated in the doctrine that the Supreme Being had assigned duties to each of the heavenly bodies by which they became rulers of the world, and that a knowledge of the Divine Will would be acquired by watching and observing the order of their motions and the recurrence of times and seasons. With the Hindus the study of Astronomy became a sacred duty, at least among the most educated classes, inasmuch as the celestial bodies were viewed as gods, and worship of them was enjoined by the Vedas. Thus the piety of the Hindus primitive ages led them to watch with care all the phenomena of the Heavens, and to perfect their calendar of festivals, etc., and to this end the first Hindu astronomers directed their attention. Thus according to Baily, accurate astronomical observations had been in India, probably before 3000 B.C., a conclusion which is justified on independent evidence. Some writers of the history of the Vedic age have also held that the sacrificial rites described in the Vedas were of astronomical origin. In any case as they were regulated by the position of the moon with reference to the stars they must be held to pre-suppose astronomical observations, which indeed became a religious necessity, so that it is reasonable to argue that an extensive astronomical knowledge was obtained in India even in the Vedic times. From the translations of Vedic hymns by the great Vedic scholars of India, we understand that even in the Vedic times the Hindus had a knowledge of the motions of the planets (at least five) and the causes of solar and lunar eclipses; and an astronomical interpretation of the Vedic hymns attempted by other scholars goes to show that a knowledge of solstitial and equinoctial points on the part of Vedic writers could reasonably be accepted.

In that early age when perfect instruments had not been invented, it was a most remarkable achievement for Hindu astronomers to have determined the lunar and solar eclipses with such correctness as would have stood the strictest scrutiny of scientific investigations. The Hindus were at a very early date well acquainted with the facts of the eclipses. They had rules for calculation of the various phases both of lunar and solar eclipses, the times of beginning, middle and end; and notwithstanding the fact that the superstitions of the people in general had greatly deferred the progress of astronomical investigations in early times, the cause of the eclipses was well understood by the Hindu astronomers as is shown by the following extracts from the *Siddhanta Siromani,* ch. 8, 1-6:—

"The moon, moving like a cloud in a lower sphere, overtakes the sun; hence it happens that the western side of the sun's disc is first obscured and that the eastern side is the last part relieved from the moon's dark body and to some places the sun is eclipsed and to others it is not eclipsed.

At the change of the moon, it often happens that an observer, placed at the centre of the earth, would find the sun, when far from the zenith, obscured by the intervening body of the moon; whilst another observer on the surface of the earth will not, at the same time, find him to be so obscured, as the moon will appear to him to be depressed from the line of vision extending from his eye to the sun. Hence arises the necessity for the correction of parallax in celestial longitude and latitude in solar eclipses, in consequence of the difference of the distance of the sun and the moon.

When the sun and the moon are in opposition, the earth's shadow envelopes the moon in darkness. As the moon is actually enveloped in darkness, its eclipse is equally seen by every one on the earth's surface, and as the earth's shadow and the moon which enters it are at the same distance from the earth, there is, therefore, no call for the correction of the parallax in a lunar eclipse.

As the moon moving eastward enters the dark shadow of the earth, its eastern side is first of all involved in obscurity and its western side is the last portion of its disc which emerges from darkness as it advances in its course.

As the sun is a body of vast size, and the earth insignificantly small in comparison, the shadow cast by the sun from the earth is, therefore, of a .conical form, terminating in a sharp point. It extends to a distance considerably beyond that of the moon's orbit.

The length of the earth's shadow and its breadth at the part traversed by the moon may be easily found by proportion".

From the above extracts we see that the correction of parallax in latitude and longitude is of great importance in calculating eclipses accurately, and hence we are led to believe that the Hindus had a thorough knowledge of this phenomenon even in the Vedic age when eclipses were correctly calculated for religious purposes.

The subject of eclipses was of a nature too sacred to be treated lightly or communicated indiscriminately, and a warning is given in the *Surya Siddhanta* to this effect—"This science secret even to the gods is not to be given to any body but to the well examined pupil who has attended the whole year." The same feeling of awe was even current amongst the ancient Chinese who looked upon an eclipse as foretelling imminent calamities. Thus going back to Chinese records, we are told, that in 2159 B.C., the Royal astronomers Hi and Ho failing to predict an eclipse were executed, so that apparently an eclipse was then regarded as an event of serious portent. This was by no means strange. For if celestial phenomena, in general, excited wonder and a spirit of worship, an eclipse could not but be associated with a temporary deluge, and would as such be naturally regarded with a feeling of mysterious awe. And even as the motion of the sun and the moon produced obvious terrestrial phenomena, day and night and changes in the seasons, so there was nothing to indicate that these motions and the positions of stars, as well as such naturally striking phenomena as the eclipses and the appearances of comets and meteors were not capable of producing occult and baneful influence on man. Accordingly it should not sound in any way strange that the people who studied these phenomena were popularly credited with and gradually came to claim supernatural wisdom. Thus in the infancy of this ancient science, astrology and astronomy became blended together in an inseparable way. From the above account of the execution of the Royal astronomers Hi and Ho we may naturally conclude that the Chinese even at that time possessed some rules of predicting the eclipses, the knowledge of which the Hindus also could claim even in Vedic times.

We should not in this connection forget to mention the superior astronomical knowledge of the Chaldeans who had apparently made a wonderful progress in astronomy; long before the Greek civilisation came into being, the Chaldeans had discovered the Saros (literally means repetition). This consists of 223 lunations in a period of about 18 years 11 days constituting a cycle and includes an exact number of periods of the revolutions of the moon's nodes, relative to the earth; and the Chaldeans had found by means of observations that eclipses during one cycle are repeated during the following cycles, exactly in the same order and almost under similar circumstances. In order to understand this, it is necessary to remember that an eclipse takes place, when the sun, the earth, the moon and the node of its orbit are almost in the same line. It will be easily seen that all the configurations satisfying this condition for an eclipse will continually recur in a fixed sequence, at every succeeding Saros on account of its peculiar characteristics mentioned above. It may be possible, and it was perhaps so, that the explanation of the remarkable property of the Saros was not known to the Chaldeans. It was probably derived as a generalisation from observations alone. In order to arrive at this generalisation, it was obviously necessary to observe and tabulate the eclipses for a considerable length of time and with an accuracy which presupposed a high standard of astronomical knowledge. In fact, the discovery of the Saros was one of the greatest achievements of ancient astronomy and it is no small credit for the Chaldeans to be associated with one of the most brilliant records of scientific generalisation.

The next step of astronomical progress leads us to the discovery of the divisions of the ecliptic and the signs of the zodiac. The question naturally arises, to whom the credit of this important discovery is due. The Hindu astronomers had two systems of reckoning: the lunar mansions or the *tithis* and the signs of the zodiac or the *rasis*, the first being obviously the earlier of the two. For while the moon's motion among the stars is a matter of direct observation, the solar motion in its relation to the stars could only be observed by an indirect method on account of the fact that its light shuts out of view all stars in its neighbourhood. On the other hand the moon's motion is much more irregular than that of the sun. The observation of the sun's motion, therefore, came to be recognised as a matter of practical as well as of scientific importance and the method of signs or *rasis* ultimately superseded the method of the *tithis*. As regards the Lunar system of the Hindus, its high antiquity is testified to by the fact that the primitive series opened with Krittika (Pleiades) as the sign of the vernal equinox. This could take place only about 2300 B.C. Nowhere else would be found a well-authenticated zodiacal sequence of so early a date. If this be granted, it seems to be very probable that the method of signs was built up in India, for the method of *tithis* which is admitted to be peculiar in India may be regarded as the parent of the method of signs and we are thus able almost to trace a gradual evolution along lines well recognised in science of the system of signs. The Hindu astronomers divided the ecliptic and the zodiac into 28 parts (and then into 27) forming so many groups of stars in the path of the moon, each division corresponding with the space of the moon's daily motion through them. The groups were hence called lunar asterisms. The ancient Hindu astronomers chose a set of 27 principal stars, one for each of the 27 lunar constellations, generally the brightest star of the asterism, and called it *Yoga-tara* whilst the cluster of stars was named the *Naksatra*. The Yoga-tara was connected with the beginning or the first point on the ecliptic of the division representing this space of the asterism by the small arc of apparent difference of longitude between them, this arc being called *Bhoga* of the asterism. Thus the 27 divisions of the ecliptic became as fixed in position as the stars themselves like a fixed dial, with the numbers ranging not along the equator but along the ecliptic itself. The accompanying diagram may, perhaps, more explicitly convey the nature of the Hindu ecliptic, which is here shown as a great circle in perspective.

Each division represents one twenty-seventh part of the ecliptic, and each star the Yoga-tara of the lunar astersim to which it belongs. From the above arguments and from the investigations of Weber and Colebrooke we deduce that as a later date the diligence and genius of the Hindu astronomers gave such an impetus to the progress of astronomical knowledge in India that it secured for them the superiority over all other ancient nations. Perhaps the rudiments of astronomy the Hindus, the Chinese, and the Egyptians gathered .from their common central Aryan home. For, recording to Colebrooke, the emigrating tribes, who undoubtedly went from, their common Asiatic home, carried with them the evidences of the common origin of their astronomies, and indirect method on account of the fact that its light shuts out of view all stars in its neighbourhood. On the other hand the moon's motion is much more irregular than that of the sun.

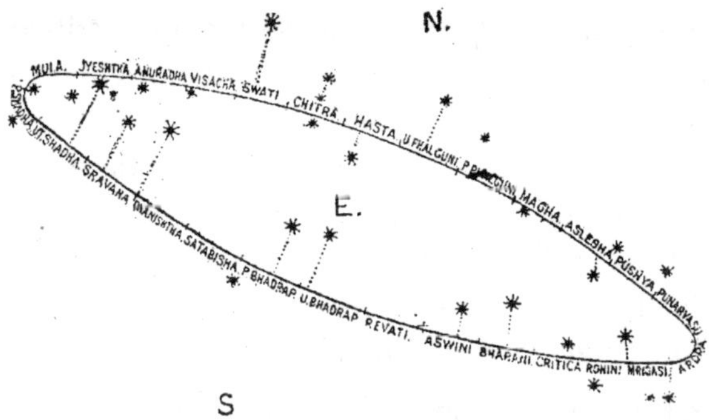

I. H.Q. December, 1925.

The Yoga-tara was connected with the beginning or the first point on the ecliptic of the division representing this space of the asterism by the small arc of apparent difference of longitude between them, this arc being called *Bhoga* of the asterism. Thus the 27 divisions of the ecliptic became as fixed in position as the stars themselves like a fixed dial, with the numbers ranging not along the equator but along the ecliptic itself. The accompanying diagram may, perhaps, more explicitly convey the nature of the Hindu ecliptic, which is here shown as a great circle in perspective.

Each division represents one twenty-seventh part of the ecliptic, and each star the Yoga-tara of the lunar astersim to which it belongs. From the above arguments and from the investigations of Weber and Colebrooke we deduce that the Hindus founded their lunar mansions which the Arabs (under the name of Manzil) and the Chinese (under the name of Sien) borrowed. We also know that the Greeks were not possessed of the system of lunar mansions. Weber, however, propounded the view that the system originated in Babylon. Such a view can no longer be maintained, since we have now to admit that the Babylonian system is based on the sun's motion. Now if the view advocated above that the lunar system must be of older date than that based on the sun's motion be correct, we have to admit that the Babylonian system was derived by adoption. Hence we are justified in inferring that if the Hindus can claim to be the originators of the system of lunar mansions, they have an equal right to claim to be the pro-pounders of the system dependent on the divisions of the Zodiac; for it is natural to believe, that a theory which belongs to a particular country in its first stage would very likely belong to that country in its second stage as a natural consequence of evolution.

But a careful examination of the scientific facts hitherto available leads us to the conclusion that the Hindu, the Chinese and the Egyptian astronomy grew up side by side, till at a later date the diligence and genius of the Hindu astronomers gave such an impetus to the progress of astronomical knowledge in India that it secured for them the superiority over all other ancient nations. Perhaps the rudiments of astronomy the Hindus, the Chinese, and the Egyptians gathered from their common central Aryan home. For, according to Colebrooke, the emigrating tribes, who undoubtedly went from their common Asiatic home, carried with them the evidences of the common origin of their astronomies, and the arguments in support of this proposition can be summarised as follows:—

(1) They had a like number of days of the week and a like number of months.

(2) They had similiar divisions of the ecliptic.

(3) They had the same signs of the zodiac.

(4) They had similiar months of the year.

(5) They had a like number of lunar constellations.

(6) They made a like use of the celestial sphere.

(7) They made a like use of the Gnomon and

(8) They had a like fantastic nomenclature of constellations.

Some are however inclined to include Greek astronomy in the above list. But a careful investigation of certain facts does surely raise some doubts as to the date of origin of the Greek astronomy which, we are persuaded to believe, cannot be placed as early as that of the Hindu astronomy, nay, the Greek astronomy was of a much later origin. For we know that it was Thales who first introduced in Greece the study of astronomy on a scientific basis and he in his turn had been taught his rudiments by the Chaldean priests. Before his time scientific investigations of astronomical phenomena had not been encouraged and it was not till a much later date that the study of astronomy was taken up in Greece on a purely scientific basis. Even up to the time of Aristotle we find no plausible record of scientific generalisations of astronomical phenomena formulated in Greece. To prove that the earth is spherical, Aristotle asserts that the sphere is the most perfect shape and the earth being the handicraft of the greatest workman must be spherical as a perfect specimen of the noblest workmanship; again in connection with the diurnal motion, Aristotle argues that the motion of the heavens is towards the right because this is the more honourable direction. Such arguments, we dare say, can hardly be called scientific reasonings, not to speak of conclusive proofs. The two great names that stand out pre-eminently in the history of Greek astronomy are Hipparchus and Ptolemy. In fact it was due to them alone that the Greek astronomy was rebuilt and remodelled on a novel scientific plan. It was in the second century B.C. that Hipparchus determined that the sidereal time taken by the sun to start from one sign and to return to the same again is less in the next year than it was in the previous year. This earlier return is termed precession. On account of this precession we get the tropical year which is the interval that elapses between the sun's starting from one sign and its coming back to the same again; whereas the sidereal year is the interval elapsing between the sun's moving from one star and reaching it back. Hipparchus calculated the respective periods of these two years, the number of days in a month and also the periodic times of the five planets then known. Besides it was due to him also that the inclinations of the Solar and the Lunar orbits with the equator were finally determined. Of course, he was indebted in no small degree to the Chaldeans for his calculations and assertions; yet it was he who first made Greek astronomy stand on a sound mathematical basis. About four centuries later came Ptolemy. Within this interval Greek astronomy made no noteworthy progress. Ptolemy was undoubtedly the most eminent of ancient Greek astronomers. His chief credit lay in compiling his greatest work *Almaghast* in which were recorded in a systematic way the notable discoveries of his predecessors and of himself. He commanded a unique influence on the public mind of his time. He asserted that the earth was fixed and the planets of the solar system were moving round the earth in circles. Of course this view was well grounded on common experience. In this connection, it will be interesting to cite the scientific reasonings which he put forward in support of his view. He said that the stars and the planets were of the nature of fire, and the earth was a collection of hard substance; hence it was more reasonable to ascribe motion to the planets then to the earth. He further added that if the earth was in motion, why should men be unconscious of the fact. But about this time Aryabhatta maintained [in the beginning of the Christian era, (see Brennand's *Hindu Astronomy)*, or 5th century A.D., (Kern and Colebrooke), the diurnal motion of the earth round its axis. "The starry sphere", he affirms, "is stationary and the earth making a revolution produces the daily rising and setting of stars and planets." The commentator, Prthudaka Svamin quotes the following lines in support of the above view:

Bhupanjarah sthiro bhur evavrtyavrtya pratidaivasikau/
Udayastamayau sampadayati naksatragrahanam//

The idea of heliacal rising and setting of stars and planets was first fully developed in Europe by Copernicus in the 15th century, and before him it was not recognised, though some rudimentary hints were thrown by Pythagoras in the 5th century B. C. Aryabhatta promulgated the above theory in its present form in the beginning of the Christian era or 5th century A. D. at the latest. We are therefore inclined to believe that this theory had its origin in India and was introduced into Europe through the Greek medium, where it was given its present practical garb. After Ptolemy the study of astronomy became obsolete in Greece and when the Arabs came as a conquering nation to settle in Europe, they grew fascinated by this ancient science and took up its study seriously. At first a few Greek treatises were translated, but soon by the untiring energy and exertions of Al-Badauni Abul Waza and others there came a new spirit of astronomical researches among the Arabs and as the result a few theories on precession and the lunar orbit were discovered. However, the above arguments lead us to believe that the Hindu, Chinese and Babylonian systems grew up side by side and under their fostering influence was developed the astronomy of the Greeks and the Arabs.

The fixing of the first point of the Indian Zodiac so as to make it unchanged in after time is characterised as a remarkable event in the history of Hindu astronomy and is one of the fundamental differences of the Hindu and the European systems. In the European astronomy all longitudes are measured by arcs of the ecliptic, whose origin is the equinoctial point at the time of observation. This point moves backward along the ecliptic at an annual rate of about 50" causing an annual change of the same amount in the longitude of the stars and this is termed the precession of the equinoctial point. Hindu astromony, however, avoided the annual change of longitude by assuming a fixed point of the ecliptic as the beginning of their system, the position of the lunar asterisms being all fixed in relation to that beginning. The Hindu astronomers were at a very early date acquainted with the phenomenon of precession. Even the theory of the Libration of the Equinoxes was early known to them. Bhaskara says—"The intersection of the equinoctial and ecliptic circles is the *Krantipata* or inter-secting point of the sun's path. Its revolutions are retrograde, three; myriads in a Kalpa". A corresponding passage occurs in the *Surya Siddhanta,*—"The circle of Asterisms moves eastward 30 score (or 600 times) in a Yuga that is to say all the asterisms at first move west ward 27", then returning from that limit they reach their former places, then they move eastward the same number of degrees and returning thence come to their own places. Thus they complete one libration or one revolution as it is called. In this way the number of revolutions in a Yuga is 600, which answers to 600000 in a Kalpa" (Vasudeva Sastri).

We shall now return to our discussion of the Zodiacal system. In modern times, with our fixed observatories, our instruments of precision, clocks and instruments, it is rather an easy affair to determine the position of the sun or any other celestial body, at any time. We have, in fact, only to note the moment of the transit of the body across the meridian of a place, by means of the sidereal clock. This gives one of the co-ordinates and the other co-ordinate is given by the altitude of the body at its meridian passage and by this means the position of the body and its motion at any time may be accurately represented. But the ancient astronomers had no such means at their disposal. They early recognised that the various groups of stars or constellations seemed to be bound together by an invisible chain and to be apparently fixed or practically so to the celestial dome or vault, which, appeared to rotate about a certain definite axis practically fixed in space. Of this rotation of the celestial vault the *Surya Siddhanta* says:—

Bhacakram dhruvayor baddham aksiptam pravahanilaih/
Paryetyajasram tannaddha grahakaksa yathakramam//

"The starry sphere is said to revolve constantly through the influence of the pravaha winds, as also do the planets confined within their respective orbits". The ancient astronomers must have noted, moreover, in the next place that there is one family of constellations, arranged along the whole of the celestial region, through which the sun, the moon and the planets (known to them) pursued their courses. The family constellations, therefore, came to be used as so many signs for indicating and describing the positions of these at any time and their motions from day to day. To explain this Zodiacal system in the light of common experience we suppose a big clock to be fixed in space with twelve divisions marked on it and the two hands indicating specified times, and then we one even at the first sight that such a clock bears a close analogy to the Zodiacal system with its twelve signs. Just as we can find the exact time simply by looking at the clock, in a similar way the Zodiacal system gives the precise position of the sun at any time even on a moment's observation. For this reason, we must admit that whoever the first inventors of the Zodiacal systems may have been, this device of the signs and asterisms seems to be of very remote antiquity and speaks volumes for the ingenuity of the early students of astronomical science. We know that the sun's path on the celestial vault is a circle. If this is suitably divided into twelve parts, each arc will be found to be occupied by a group of stars, called a sign of the Zodiac irregularly placed but in such a way that the group may be taken roughly to give a distinctive character to the particular sub-division which it occupies. Starting with any time of reckoning from an equinox (that is, the moment at which the sun is at the equinox or when we have equal day and night, throughout the earth) each of these signs will be passed over roughly in one month and one mode of describing the sun's motion would obviously be to name the particular sign and the position in that sign that the sun occupies, at any particular epoch. In the same way, the lunar path in the celestial vault-being also a circle, this path might also be used in the same way as the circle of reference. This latter circle was used by the Hindus, who divided its circumference into twenty seven parts. Now as the sun's path as well as the moon's are contained within the same belt of the celestial vault, the solar as well as the lunar positions might be described with reference to the *tithis* as well as to the signs. Also the fact that the motions of the sun, the moon and the planets are all confined to a narrow belt with the ecliptic as the central line enhanced the usefulness of these methods of representation. The lunar month consists of 30 lunar days or *tithis.* It is the moon's synodic period from one new moon to the next, and the thirtieth part of this period is, therefore, the lunar day. A solar day is the time which the sun takes to move from the beginning of one sign of the Zodiac to the next. A solar year consists of 12 solar months, and this is called a Divya Day or a day of the gods (See *Surya Siddhanta).*

Thus when the motion of the sun and the moon became completely known and their positions accurately predicted, the determination of the eclipses was naturally the next stage in the development of observational astronomy. This calculation of the eclipses the ancient astronomers performed with great skill and the solution was very correct, though naturally not so accurate as modern methods will yield. For it should be noted that for a successful solution of the problem of an eclipse, it was not necessary to definitely grasp the fact that it was the earth that was in motion; the result could have been the same, if the earth were at rest and the sun was moving about it, as the phenomenon is dependent on the motion of the cone of shadow cast by the earth, relatively to the moon and its node, and this motion would be the same, whether the sun is at rest and the earth is in motion, or the contrary is true. Thus in the *Surya, Siddhanta;*— "The earth's shadow is always six signs from the sun. When the place of the moon's node is some degrees within or beyond the place of the shadow, the same thing will take place. At the end of a new-moon, the sun and the moon are of the same signs and at the end of a full-moon, they are at a distance of six signs. The moon, being like a cloud in a lower sphere, covers the sun in a solar eclipse; but in a lunar eclipse the moon moving eastward enters the earth's shadow, and the shadow obscures her disc."

This observation of the sun's motion with reference to the signs of the Zodiac must have very early led to the discovery of the phenomena, namely that at each succeeding equinox, the sun does not come to the same stars, but that the signs and the stars are observed to have a motion relative to the point, which the sun occupies at either equinox and that the direction of motion is opposite to the sun's observed annual motion among the stars. In 134 B.C. Hipparchus discovered this fact on observing a star which was new to him, but the precession was apparently long known to Hindu astronomers, and its rate determined by them roughly to a near approximation.

Now it was a very remarkable achievement for the ancient astronomers; for the discovery of precession was essential to the progress of accurate observational astronomy. Hence we shall put in here a few words to explain this phenomenon by assuming results that belong to modern astronomy, following the method of synthesis as far as practicable. We know that the path of the sun in the celestial vault is accurately a circle and it follows that its orbit must be a closed plane curve. An observation of the stars which may be regarded as fixed to the celestial vault and in space leads to the conclusion that this plane is fixed in space. The line perpendicular to this plane through the centre of the celestial vault is, therefore, fixed in direction in space and precession consists in the rotation of the earth's axis about this line in a period of about 26000 years. The point at which the polar axis meets the celestial vault thus describes a small circle in space as a necessary consequence, the stars that occupy the region marked by this circle become pole stars in succession. While this goes on, the line of intersection of the equator and the ecliptic (which passes through the sun at an equinox) points to different stars at different epochs. This apparent change in the position of the stars is called libration and the polar axis is called its fulcrum. We have in the *Surya Siddhanta:*— "The circle of asterisms librates 600 times in a great Yuga, the libration being performed through 27° from a mean point in a retrograde direction; then returning to the mean point, proceeding through 54° with a direct motion, thence returning to the mean point with a retrograde motion, thus making 108° for a complete libration". The rule further directs the calculation to be carried backwards proportionally, the quantity of it to be computed by means of the days in a Kalpa or a Yuga, and the days which have elapsed from the assumed epoch. When the sun is at the equinoxes and at the solstitial points, its observed position give the amount of libration or the amount of precession of the equinoxes. If the true longitude of the Sun be less than the observed longitude, the asterism librates towards the east by the amount of difference and if greater, the asterism librates towards the west by the amount of excess.

The motion of the earth to which precession is due is a phenomenon well-illustrated by the common peg-top, though in this case we are dealing not with rotation about the centre of gravity of a body free in space, but with rotation about a fixed point. While the top is in rapid rotation about its axis of symmetry, the axis is itself slowly describing a cone round the vertical. Thus the axis of the peg-top is at each instance moving in a direction at right angles to that in which the force of gravity would appear to urge it and which it is only prevented from following by the fact that the top has a rotation about its axis very much more rapid than the conical motion of the axis itself. The axis of the top will here represent the earth's axis, the vertical, the axis of the ecliptic, and the spin the motion of the earth (to which the apparent diurnal motion of heavenly bodies is due).

It is this precession, which among other causes led to a great confusion in calculating the months of the calendar. Defining in general terms the period in which the Sun completes its cycle as the year, it is easily seen that the cycle may be said to be completed, either when the Sun returns to the same point in space, the same point of a *tithi* (or a *rasi*) or to the same solstice (or same equinox) as the one from which the year is reckoned. The first is called the tropical year and the second the sidereal year. Again the year may be determined by lunar months or the periodic times of other celestial bodies, as has been described in the *Surya Siddhanta*. Further the difficulty is enhanced by the fact that none of these periods contain an exact number of days and the solar day is not a constant interval time; hence it is a matter of no little wonder how the problem of the calendar was so successfully solved in India, China, Chaldea, Egypt and Greece in ancient times. The next point that engages our attention in the gradual development of ancient astronomy is the explanation of the planetary motions and the method of epicycles. The first notion of planetary motions was stated thus in the *Surya Siddhanta:* "The planets in their orbits go rapidly and continually with the stars towards the west, and hang down at an equal distance as if overpowered by the stars. Therefore the motion of the planets appears towards the east, and their daily motions, determined by their revolutions, are unequal to each other in consequence of the circumferences of their orbits; and by this unequal motion they pass the sign. The planet, which moves rapidly, requires a short time to pass the sign, and the planet that moves slowly passes the sign in a long time". This opinion that the motion of all the planets was caused by a velocity in their orbits, which was the same for all alike, was prevalent not only in the east but also in Europe even in the times of Kepler and Newton. This is evident from the manner in which Kepler combated this doctrine and the important use he made of it. Soon after the death of Tycho, Kepler made many discoveries from Tycho's observations. He found that astronomers had erred from the first use of science in ascribing circular orbits to the planets. He easily saw that the higher planets not only moved in greater circles, but also moved more slowly than the nearer ones, so that on a double account their periodic times were greater.

This planetary motion the Hindus and the Greeks explained by means of epicycles. Aryabhatta ascribed to the epicycles, by which motion of a planet is represented, a form varying from a circle and nearly elliptic. It was known to the Hindus that a supposed uniform motion in a circle about the earth did not really represent the true motion in its orbit, although the hypothesis served sufficiently to determine the mean motions and the mean place of a planet when deduced from observations carried on for lengthened periods. They knew that every planet in its course was subject to great irregularities, the motion undergoing continual changes. At one time it would be direct towards the east until the planet reached a stationary point where it would seem to be at rest; then a retrograde motion would ensue and continue till another stationary point was reached, and the eastward motion could be repeated. It was to account for these inequalities that the Epicycle was invented. By the Greeks this contrivance was ascribed to Apollonius. He conceived that a planet in its course described, with uniform motion, the circumference of a circle, called the Epicycle, whose centre *moved uniformly in the circumference of another circle* called the deferent, the centre of which was the centre of the earth. It was also supposed that while the centre of the Epicycle was moving eastward in the direction of the signs, the planet itself was moving in a direction contrary to that of the signs. By this hypothesis it was easy to show the various changes in the motions of the planets. This theory was generally adopted by the western nations with the addition of Epicycles, introduced by Ptolemy, as necessary for expressing the apparent motions with accuracy. But the Hindus had two different methods for calculating "the true place" of a planet from its mean place. One of these methods resembled that of Apollonius, but there was considerable difference: that while the planet moved in the deferent concentric with the earth, *the Eyicycle itself was conceived to be variable, the circumference being greatest when the planet was in an apsis (at Apogee or Peragee, the true and mean places being then coincident) and least when the planet was at a distance of 90° from the points.* The other method supposes that while the mean place of a planet is a point moving uniformly eastward round the circumference of a circle whose centre is the earth, the Planet also moves uniformly eastward, in the same time round the circumference of an equal but eccentric circle whose centre is situated in the line joining the Apogee with the centre of the earth, the distance from it being the eccentricity. These two methods, whether by assuming the motions as being in an eccentric or in an Epicycle, give exactly the same results. These main points of the Greek and Hindu Epicyclical theories show that unlike the Epicycles of Ptolemy and other Greek Astronomers, the Indian Epicycles had a variable circumference, that of the first Epicycle being largest at Apogee and Perigee, varying from those points through the deferent to its places at the quadrants where its circumferences were least.

Here we give in a few words a short review of ancient astronomy. We have tried to follow in a synthetic method the gradual developments of this ancient science and we are led to believe that its progress was remarkable inasmuch as it paved the way for further scientific generalisations. Even it is conceivable that the principle of gravitation in its elementary form was known to the early Hindu writers of astronomy. The following occurs in the second chapter of *Surya Siddhanta*: "The solar disc being heavier, the force of attraction on it is less and the lunar disc being lighter the force of attraction on it is greater".

In fact, we think that this passage has a special significance with regard to the principle of gravitation, for the first part of Newton's theory states that the forces of attraction are directly proportional to the masses. Even Varahamihira writes in the sixth century: "The earth attracts that which is upon her." It is also given by Brahmagupta, in the following more complete form, "All heavy things fall down to the earth by a law of nature, for it is the nature of water to flow, that of the fire to burn, and that of the wind to set in motion." Or as in *Siddhanta Siromani:* "The earth has the power of attraction and therefore it attracts heavy things towards it; the general notion is that of falling". But one thing strikes us most when engaged in the study of ancient Astronomy. In many places we notice that the ultimate generalisations are only recorded but the intermediate links are lost. At least this is often noticeable in the case of Hindu astronomy. We have there many remarkable scientific generalisations noted down but the method of procedure that must have been required to arrive at these generalisations is not recorded. This probably led some oriental scholars to believe that those theories are of foreign origin. But we think there is a sufficient reason for this omission. In India, we know, there prevailed the custom of oral transmission of knowledge in literature as also in science. When a preceptor arrived at any general result he would have it recorded in the form of a sloka; he would of course explain the method of procedure to his pupils and ask them to get the sloka by heart. Thus with the succession of preceptors as years rolled on, the number of slokas became so large as to demand a systemetic collection of them, for a compilation served the purpose of constant reference. The pupils had this collection from their guru and would learn the processes orally from him. But certainly there came a time when the study of science was neglected in India. The result was that the methods of procedure which had been regarded as the connecting links were lost and only the general theories survived. Thus perhaps originated the greatest treatise of astronomy of the Hindus, namely, the *Surya Siddhanta*.

However, when all is said, this should not be denied that in the study of astronomical facts and in the observation of astronomical phenomena the early thinkers brought with their wonderful power of investigation a noble spirit of enquiry and a love for the truth—a love almost akin to a spirit of veneration for the grand and the sublime. Thus when bewildered with the awful mystery of the heavens or the rich panorama of the nocturnal sky, they would in a fit of momentary trance exclaim:—"O Lord, O omniscient Creator, remove my doubt, for none else save Thee can fathom this deep mystery !"

Sukumar Ranjan Das

8
A Short Chronology of Indian Astronomy

The ancient Indian astronomy had for its basis the religious ceremonies of the Hindus. In the early Vedic period each heavenly body represented a Divinity and so the study of astronomy originated in the doctrine that the Supreme Being had assigned to the heavenly bodies some duties by which they became rulers of the affairs of the world, and that a knowledge of the Divine Will could be had by observing the order of their motions and the recurrence of times and seasons. The Hindus therefore watched with care all the phenomena of the heavens in order to perfect their calendar of festivals etc. The Vedic sacrificial rites had a close connection with astronomical calculations, and so some amount of astronomical knowledge was obtained in India at the time when the Vedas and the Brahmans were composed. It is from astronomical premises that scholars like Tilak, Diksit and Jacobi argue a great antiquity for the vedic writings.

The vedic year was twelve months of 30 days each, with an occasional intercalary month, "the thirteenth month fabricated of days and nights, having thirty members." (Atharva Veda xiii, 3, 8). There is no indication of any definite cycle, the five year cycle having appeared later. The year was also divided into two equal courses or ayanas, a northern course or Uttarayana beginning at the winter solstice and a southern course or Daksinayana beginning at the summer solstice. In the Rg-veda two asterisms only are mentioned, viz., Magha and Phalguni; but in later vedic texts (A. V. xix, 7, 1-5) a complete list of the twenty-eight naksatras or asterisms is given. The list is headed by Kritikas or the Pleiades, which marked the vernal equinox of the vedic year. Jacobi proved that the vedic year commenced with the summer solstice. His arguments are based on the following rendering of a verse of the "Frog Hymn":—

"Those leaders of rites observe the institutes of the gods, and disregard not the season of the twelfth month as the year revolves and the rains return, then scorched and heated they obtain freedom."

Diksit,[1] from a passage of the Brahmanas,[2] fixes the age of the Vedas at 3000 B.C. The words "they (the Kritikas) do not move from the eastern quarter while the other asterisms do move from the eastern quarter" are taken by him to mean, definitely, that the asterism Kritikas (Pleiades), and no other asterism, was on the equator. He writes, "in my opinion the statement conclusively proves that the passage was composed not later than 3000 B.C." In vedic texts no definite mention is made of the planets, but some oriental scholars hold that the people of the Vedic age[3] had knowledge of the planets. Vedic astronomical facts are not accurate as it was not the intention of the Vedic writers to deal with astronomy directly and that it was only when religious rites demanded it, that they referred to astronomical facts.

The earliest formal Hindu astronomical work was the Jyotisa Vedanga. It is the name of certain works or classes of works, regarded as an auxiliary to the Vedas. The Jyotisa Vedanga[4] is a small metrical treatise containing a statement regarding the place of the winter solstice at some ancient date which gave rise to a good deal of comment and speculation. Its author intended to propound the doctrine of the cycle consisting of five years of 1830 apparent solar days. The year was tropical and began with the white half of the month Magha and terminated with the dark half of the month Pausa (verse 5). The year which is the fifth part of the yuga contains three hundred and sixty-six days, six seasons, two ayanas (the northern and southern progress of the sun), twelve solar months (verse 28); the sun accomplishes five tropical revolutions, the moon accomplishes sixty-two synodical and sixty seven periodical revolutions and the whole period comprises sixty-one savana months of 30 natural days each (verse 31); the yuga begins with the winter solstice at the new moon of Magha, the new moon taking place in the point of Sravistha. This indicates that the Vedanga Jyotisa was composed in the twelfth century B.C., a conclusion which has also been confirmed by the internal evidence from Bodhayana Srauta sutra. The Vedanga Jyotisa does not teach much of astronomy and it does not claim to do so. Max Muller remarks in this connection that "the object of this small tract is not to teach astronomy. It has a practical object which is to convey such knowledge of the heavenly bodies as is necessary for fixing the days and hours of the Vedic sacrifice."[5]

Jyotisa Vedanga is followed by the astronomy of the Jainas. The only work on Jaina astronomy, now available, is Suryaprajnapti. There is however evidence that two more works on Jaina astronomy were written, one called Candraprajnapti and the other Bhadrabahaviya Samhita of Bhadrabahu. These two works are mainly known from quotations by later astronomers. Jacobi in his preface to the translation of the Kalpasutra, remarks that the Bhadrabahaviya Samhita was mainly an astrological work. Bhadrabahu was the religious preceptor of Candragupta and died in 298 B.C. Bhadrabahu wrote a commentary on Suryaprajnapti and this is believed to be the first commentary on the work. The exact date of composition of Suryaprajnapti is not known, Thibaut says that this book must have been written before the Greeks came to India, as there is no trace of Greek influence in this work. Its authorship is attributed to Mahavira. The traditional date of Mahavira's death is 527 B.C., but Jacobi says that Mahavira died in 462 B.C. Therefore, it is certain that Suryaprajnapti must have been written in about 500 B.C. Weber points out that the doctrine propounded in the Suryaprajnapti shows in many points an unmistakable resemblance with that contained in the Jyotisa Vedanga and it thus became manifest that the astronomical books of the Jainas do not only furnish information about the opinions held by a religious sect, but may, if rightly interrogated, yield valuable material for the general history of Indian ideas. According to the Suryaprajnapti, the *yuga* begins with the summer solstice, at the moment when the moon is full in the beginning of Abhijit and the sun consequently stands at Pusya. The distinctive doctrine of the Jainas is the assumption of two different suns, two moons and two sets of constellations. In this connection, Colebrooke remarks,[6] "they (the Jainas) conceive the setting and rising of stars and planets to be caused by the mountain Sumeru and suppose three times the period of a planet's appearance to be requisite for it to pass round Sumeru and return to the place where from it emerges. Accordingly they allot two suns, as many moons, and an equal number of each planet, star and constellation to Jambudvipa; and imagine that these appear on alternate days, south and north of Meru." This doctrine has been controverted by all later astronomers. Here also the yuga comprises five years of 360 days each, each year in its turn being divided into twelve months of 30 days each; in the Suryaprajnapti this kind of year—commonly known as the savana year—is called the karma year or ritu year. The six days by which this year is shorter than the solar year are called atiratras. The one important point in which Suryaprajnapti differs from the statement in the Jyotisa Vedanga is this that according to the Jyotisa Vedanga the yuga begins with the winter solstice, at the moment when it is new moon, sun and moon being in conjunction in the beginning of the naksatra Dhanistha, while according to the Suryaprajnapti the yuga begins with the summer solstice, at the moment when the moon is full in the beginning of Abhijit and the sun consequently stands at Pusya. In his paper on Suryaprajnapti, Thibaut in conclusion, draws attention to the striking resemblance which the cosmological and astronomical conceptions of an old Chinese book called Tcheon-Pei (of which a complete translation was published by E. Biot in the Journal Asiatique, 1841, pp. 592-639) bear to the ideas on the same subject as expounded in the Suryaprajnapti.

The history of Hindu Astronomy of the period from 500 B.C. (i.e., the approximate date of composition of Suryaprajnapti) to 500 A.C. (i.e. the approximate date of Aryabhata's writing) is generally called the age of Hindu Astronomy. Any work written during this period of 1000 years is not available to us. But it is not probable that this period was barren of any astronomical improvement and this science remained in a static condition in this country. References of astronomical knowledge prevailing in this period are found here and there in literary and philosophical works of that period. During this period of 1000 years probably the astronomical Samhitas and the older Siddhantas were composed. Astronomy was stated to be one of the principal accomplishments of the Jaina priest in the Bhagavati sutra.[7] The date of the Bhagavati sutra is about 300 B.C. It is noteworthy that the knowledge of astronomy is considered necessary for the Jaina priest for exactly the same purpose as it was for the Vedic priest, viz., to find the right time and place for the religious ceremonies. It is also observed by Santi-candra gana (1595 A.C.) in the preface to his commentary on the Jambudvipa prajnapti. The Buddhist monks were at first advised to refrain from the study of astronomy[8] which was condemned as a low art (tiracchana vijja) as far as bhiksus were concerned. Later on Buddha relaxed his opinion and made it a rule of conduct of the bhiksus living in the woods that they must learn elementary astronomy. The incident which compelled him to change his mind is this: Once upon a time some robbers approached certain bhiksus living in the forest and asked them, "with what constellation is the moon now in conjunction?" The bhiksus could give no reply as they were ignorant of astronomy, being forbidden by religious injunction to learn astronomy. This incensed the robbers who beat the bhiksus and went away. When Buddha came to know of this unfortunate incident, he made the following rule: "They (the bhiksus living in forests) should learn the stations of the constellations, either in whole or in parts, and they should know the directions of the sky."[9] These instances prove beyond doubt that the science of astronomy was not in a static condition but was progressive during the period of 1000 years from 500 B.C. to 500 A.C. This is the period, therefore, when the astronomical Samhitas and the older Siddhantas were written. The Jyotisa Samhitas are no longer extant, though Dr. Kern was known to have possessed a fragmentary manuscript of Garga Samhita. They are mainly known from citations by later astronomers. The authors of the two Samhitas, generally mentioned in the writings of later Hindu astronomers, are Parasara and Garga. The precise period in which they lived is a vexed question. Parasara's sayings have been quoted by Varahamihira in his Brihat Samhita and also by his commentator Bhattotpala. According to the Nirukta, Parasara was the son of Vasistha and according to the Mahabharata and the Visnupurana Parasara was the grandson of Vasistha, and Parasara, was the father of Vyasa, the compiler of the Mahabharata. Parasara, therefore, must have lived just before the Mahabharata was composed, Diksit and Jacobi from astronomical evidence have fixed the date of the Mahabharata as 450 B. C. It has, for this reason, been asserted that Parasara lived about 500 B.C. Garga, the writer of Garga Samhita whose sayings have been cited by Varahamihira in his Brihat samhita is mentioned in the Mahabharata[10] as a well-known astronomer. The description, which is given in the Mahabharata of astronomer Garga, leaves no doubt as to the identification of the writer of Garga Samhita with the Garga mentioned in the Mahabharata.[11] Therefore, Garga also lived about 500 B.C. There is mention of two other Samhitas by Devala and Kasyapa in the Sukracaradhyaya of Brihat Samhita, but these works seem to have been written much later than the Samhitas of Parasara and Garga. Next comes the works on the Siddhantas. Of the eighteen or twenty ancient astronomical works referred to by ancient Hindu writers, under the name of Siddhantas, or "Established conclusions" nine are mentioned by Abul Fazel in the Institutes of Akbar, namely:— (I) Brahma Siddhanta, (2) Surya Siddhanta, (3) Soma Siddhanta, (4) Brihaspati Siddhanta, (5) Garga Siddhanta, (6) Narada Siddhanta, (7) Parasara Siddhanta, (8) Pulastya Siddhanta and (9) Vasistha Siddhanta. Other Siddhantas mentioned are the works of (10) Vyasa, (11) Atri, (12) Kasyapa, (13) Marici, (14) Manu, (15) Angiras, (16) Lomasa, (17) Pulisa, (18) Yavana, (19) Bhrgu and (20) Cyavana. The first is supposed to have been revealed by Brahma the second by the sun, the third by the moon and the fourth by Brhaspati (Jupiter). All the other tests are supposed to have been composed by mortals, and of these few are now extant, they being principally known from citations of mathematical writers of later date. It is impossible to say with certainty which of the two principal astronomical works, the Brahma Siddhanta or the Suryasiddhanta (Saurasiddhanta), is the more ancient, though the former is supposed to have been the earlier. Both of these works have undergone revision at different periods. Brahmagupta based his Brahmasphutasiddhanta (628 A.C.) on a revised edition of Brahmasiddhanta. Narsimha, the commentator of the Surya Siddhanta, affirms that Brahmagupta's rules are framed from the Visnudharmottara Purana, which the Brahmasiddhanta is contained. Various other works of the same name are referred to as being anterior to the work of Brahmagupta; such as the Brahma Siddhanta of Sakalya, which is regarded as one of the five systems from which Varahamihira compiled his Pancasiddhantika. The five Siddhantas forming Varahamihira's Pancasiddhantika are the Paitamaha (Brahma), Vasistha, Romaka, Paulisa, and Saura Siddhantas. Varahamihira there also states his view as to their order in importance and accuracy, assigning the first place to the Surya Siddhanta, placing next the Romaka and Paulisa Siddhantas as about equally correct, and declaring the two remaining works to be greatly inferior to the three mentioned. In no case Varahamihira has obliterated the characteristic features of the Siddhantas he had to deal with, and that whatever distinguishes those works from one another in the text of the Pancasiddhantika actually distinguished them also in their original form. The Surya Siddhanta of Varahamihira is the only one of the five Siddhantas, which has come down to our time, and therefore allows us to compare it with what Varahamihira tells us about the Surya Siddhanta as known to him. A cursory survey of those chapters of the Panchasiddhantika, which treat of the Surya Siddhanta, shows at once that the treatise of that name known to Varahamihira agreed with the modern Suryasiddhanta in its fundamental features. The methods of the two treatises are essentially the same and on the other hand, sufficiently different from those of the other Siddhantas summarised by Varahamihira, to ensure to the Suryasiddhanta in its two-fold form a distinct position of its own.[12] Of the Paitamaha Siddhanta Thibaut say,[13] "that the Paitamaha Siddhanta, known to Varahamihira, represents Hindu astronomy as: not yet affected by Greek, influences, and thus belongs to the same category as the Jyotisa Vedanga, the Garga Samhita, the Suryaprajnapti and similar works. From what Varahamihira says about its contents it might be identified with the Jyotisa Vedanga." The Romaka

Siddhanta, as the name suggests, must have been an adapation from some Greek or Roman astronomical work. The author of the Romaka borrowed his fundamental period from the west; the year of the Romaka, down to seconds, is the tropical year of Hipparchus or of Ptolemy who accepted the determination by his great predecessor. Thibaut has proved that there were issued several recasts of the original Romaka Siddhanta epitomised by Varahamihira.[14] It, therefore, continued to be held in esteem which prompted its remodelling in order to make it harmonize with the altered state of opinion. But the Romaka Siddhanta was referred to by Brahmagupta in a decrying spirit and its followers were censured by him. The Paulisa Siddhanta resembles in many respects the current Siddhantas of the Hindus. Some European scholars believe that the Paulisa Siddhanta is an adaptation of some work of Paulus Alexandrinus. Dr. Kern in his preface to the Brhat Samhita says, "We have no right, whatever, to infer that Paulus Alexandrinus and Paulisa are one and the same, for identity of name is too slender a ground specially when the name happens to be a common one." Further, the Paulisa Siddhanta is an astronomical work based on mathematical calculations, whereas the work of Paulus Alexandrinus is mainly an astrological one. Thibaut says that it appears that the Paulisa Siddhanta was a work following the same general methods as the Surya Siddhanta, Aryabhata and all the later astronomers; at any rate it agreed with the great majority of Hindu astronomical, works in establishing a mahayuga which contains an integral number of Savana days, etc. and of revolution of the planets.[15] He further says that in this work the new knowledge imported from the west bodied. About the Vasistha Siddhanta the Pancasiddhantika gives only very scanty information. Varahamihira places the Vasistha Siddhanta, together with the Pitamaha Siddhanta, in the lowest rank of the works whose tenets he reproduces. Thibaut says "that the methods in the Vasistha Siddhanta are so crude and they so completely omit to distinguish between mean and true astronomical quantities that the Vasistha Siddhanta can hardly be included within scientific Hindu astronomy."

The last and the most important period of Hindu astronomy beginning with Aryabhata (b. 476 A.C.) was a real advance in mathematical astronomy. Aryabhata wrote two works on astronomy of which one—Aryabhatiya is now available. He based his work on the main principles laid down in the old Surya Siddhanta. He was profusely quoted by Brahmagupta and other subsequent astronomers. Aryabhata believed in the diurnal motion of the earth. He was also the first among Hindu astronomers to explain thoroughly the planetary motions by means of epicycles. The father of Indian epicyclic astronomy,[16] as he is called, ascribed to the epicycles, by which motion of a planet was represented, a form varying from the circle to nearly an eclipse. Next came Varahamihira who lived in the sixth century A.C. He is mainly known as a compiler. His astronomical works now in existence are Brihat Samhita and Pancasiddhantika. Brihat Samhita is an astronomical as well as an astrological work and is supposed to be based on some ancient Samhita. Pancasiddhantika belongs to the class of the so-called Karanagranthas i.e. compendious astronomical treatises which do not set forth the theory of the subject at a comparative length as the Siddhantas do, but merely supply a set of concise and often only approximately correct rules which suffice for the speedy performance of all the more important astronomical calculations.[17] Varahamihira was the only one among Hindu writers on astronomy who thought it worth while to give an exposition of all the more important forms of astronomical doctrines current at his time and he gave them in the order of importance the five Siddhantas whose teaching he summarizes are to be arranged. In the Pancasiddhantika Varahamihira remarks, "when the return of the sun took place in the middle of Aslesa, the tropic was then correct. It now takes place from

Punarvasu". The Vedanga Jyotisa made the year begin with the Winter solstice and this was in vogue at the time of Varahamihira. He introduced a change in the year-beginning and made the list of Naksatras commence with Asvini. This modified year-beginning has since then been current in India. Another astronomer was Lalla who was probably a contemporary of Varahamihira.[18] He wrote Sisyadhivrddhida, an astronomical work based on the writings of Aryabhata. Though he declares himself to be a disciple of Aryabhata, he did not subscribe to his master's theory of earth's rotation and in its refutation he said, "If the earth is moving at a very rapid speed, why does not an arrow projected upwards fall on the western side of the projection? Why do not the clouds appear to move only towards the west? You cannot say that the earth is moving at a less speed, as it has to complete one revolution round in its axis in 24 hours". Almost contemporary with Varahamihira was an astronomer called Bhaskara. He was not the great astronomer Bhaskaracarya, the writer of Siddhanta Siromani. This Bhaskara wrote two astronomical works, called Brhat Bhaskariya and Laghu Bhaskariya, based on the teachings of Aryabhata. He took 444 Saka year for the karana year and must have, therefore, lived about 500 Saka or about 578 A.C. He was at least anterior to Brahmagupta who censures Aryabhata for some of the omissions in the latter's methods. Prthudakasvamin, the commentator of Brahmagupta, remarks in this connection that Brahmagupta was not just in censuring Aryabhata as it was not his omission, but an omission of his disciples, Bhaskara and others (Bhaskaradinam bhavatu).[19] The great Bhaskaracarya also mentions some Aryabhata's disciples, naming Prabhakara and others (Aryabhatasisyah Prabhakaradayah). This Prabhakara is no other than the above named Bhaskara. The second Bhaskara did not call the first one. Bhaskara as that might lead to confusion and therefore gave the synonym Prabhakara for Bhaskara. This was also a common practice in Sanskrit works.

The next astronomer who came into prominence was Brahmagupta who wrote his famous work, Brahmasphutasiddhanta in 550 Saka or 628 A.C., when he was thirty years old. Narsimha, the commentator of Surya Siddhanta, affirms that Brahmagupta's rules are framed from the Visnudharmottra Purana, in which the Brahmasiddhanta is contained. The above assertion is also confirmed by Prthudakasvamin, Brahmagupta's commentator, and Alberuni. Brahmasphutasiddhanta was translated into Arabic by Mohammed Ibn Ibrahim Al-Fazari in 773 A.C. and the translation was known by the name of Sindhind. Another work of Brahmagupta, a Karanagrantha of the name of Khandakhadyaka, was translated into Arabic and goes by the name of Alarkand. It is necessary to mention in this connection that Brahmagupta maintained his karana to be founded on Aryabhata, or at any rate to give results equal to those to be derived from Aryabhata; but it is curious that the dimensions of the epicycles and the positions of the apogees assumed in the Khandakhadyaka differ, all of them more or less, from those recorded in the Aryabhatiya. It is hence probable that the Aryabhata whom Brahmagupta is supposed to have followed is a person entirely different from the famous Aryabhata. He cannot be Aryabhata II, the writer of Mahasiddhanta, who was much posterior to Brahmagupta. Therefore it is probable that this Aryabhata whom Brahmagupta followed must have been a third Aryabhata anterior to Brahmagupta and posterior to Aryabhata I. Brahmagupta refuted the theory of earth's rotation of Aryabhata I and put forward as an argument that "if the earth moves a minute in a prana, then whence and in what route does it proceed? If it revolves, then why do not lofty objects fall?" After Brahmagupta there was for some time dearth of eminent astronomers. The next astronomer of importance was Munjala living in the beginning of the tenth century A.C. He wrote his Karanagrantha, Laghu manasa in

932 A.C. He must have been a very celebrated astronomer as he was mentioned by name and even in some points (specially in his methods of finding the processional rate) followed by the great astronomer and mathematician Bhaskara. In the Laghu-manasa Munjala states that the annual precession was 59.9007 seconds of arc. The next astronomer was Sripati. He also wrote an astrological treatise. His astronomical works are Dhikoti, a Karnagrantha, and a Siddhanta called Siddhantasekhara. His exact date is not yet known, but this much is known that he was anterior to Bhaskaracarya and wrote his Siddhantasekhara about 962 Saka or 1040 A.C. He gave an account of the peculiar doctrine of the Jaina astronomers who assumed two suns, two moons, two sets of stars and planets and the Meru to be of pyramidial shape. Just following Sripati was Bhojaraja, another astronomer not very much known. Dr. Bhaudaji says that Bhojaraja wrote a Karanagrantha called Rajamrganka, from which it is known that Bhojaraja who was the king of Dhara, was born in 964 Saka or 1042 A.C. The next astronomer was Satananda, who was prominent and popular with the Indian calendar makers. His work, Bhasvati based on the Suryasiddhanta of Varahamihira is a karana grantha, a compendium, and is made much use of by the calendar makers. Its popularity with them is known by the saying "Bhasvatigrahane dhanyah." Bhasvati was written in 1021 Saka or 1099 A.C. But the most celebrated among Indian astronomers was Bhaskara who lived in the twelfth century A.C., having been born in 1114 A.C. At the age of 36 he wrote his great work, Siddhanta Siromani, divided into two parts, Goladhyaya and Grahaganitadhyaya. Besides this, he wrote at a much later period a Karanagrantha called Karanakutuhala, when he was 69 years old. Recently an astronomical work, a short treatise on the corrections of the moon's place called Bijopanayal has been stated as a distinct astronomical work of Bhaskara. But from a perusal of the contents one may doubt if the work can at all be attributed to

Bhaskara. There are passages in the Vasanabhasya of the work which lead to this doubt. If this work is at all written by Bhaskara, it was never meant to be a separate treatise. It must be an appendix to Bhaskara's Siddhanta Siromani to which his well known Bijaganita and Patiganita are also appendices. Pandit Bapudeva Sastri has given a brief account of Bhaskara's genius in J.A.S.B. of 1893. About Bhaskara Spottiswoode remarks—"That the penetration shown by Bhaskara in his analysis is in the highest degree remarkable; that the formula which he establishes and his method of establishing it, bear more than a mere resemblance to the corresponding process in modern mathematical astronomy; and that the majority of scientific persons will learn with surprise the existence of such a method in the writings of so distant a period and so remote a region."

In conclusion, a few words are necessary to deal with the question of the foreign influence on Indian astronomy. This is a long disputed and has been discussed thread-bare by the oriental scholars. Many scholars tacitly assume a Greek basis for the Hindu scientific astronomy. Thibaut in his Introduction to the Panca Siddhantika has discussed this question at a considerable length and concludes that "the later Siddhantas fall under one category, all of them, however much they differ in details, representing the modern phase of Hindu astronomy which is completely under the influence of Greek teaching." Alberuni, Rodest, Weber, Colebrooke, Whitney and various Arab scholars have also dealt with this question from different points of view. The general hypothesis is the dependence of Hindu scientific astronomy on Greek teachings. Various facts have since then been found against this pre-conceived hypothesis and discussed at a great length. But the fundamental belief of the pro-Greek school has not yet been shaken. Some of the Arab writers, especially Al Fazari, the writer of Sind-hind, have expressed the view that the Greek and Hindu mathematicians drew their inspiration from a common source. It is, in this connection, worth while to consider the great Hebrew cabalist (Cabala, gabbalah) compendium, the *Zahar,* passages from which have been discussed by Gunther; since it is the opinion of many Hebrew scholars that the authors of the *Zahar* were influenced by Hindu writers, this fact may throw some light upon Hindu originality in astronomy. Bishop Severus Sebokht of Nisibus (c. 650 A.C.) who would naturally have been expected to favour Greek claims, took up a quite contrary position and believed in the original and independent growth of Hindu Spherical astronomy. Since he was almost a contemporary of Brahmagupta, his testimony might have some special significance regarding this question of foreign influence on Indian astronomy. However, the final seal has not yet been put and it will require further researches on this subject to arrive at a conclusive decision regarding this question of Greek influence.

References

1. *Indian Antiquary,* 1895, xxiv, p. 245.
2. *Satapatha Brahmana,* II, 1, 2. 2-4.
3. *Bharatiya Jyotih Sastra* by S. B. Diksit; *Our Astronomy and Astronomers* by Jogesh Chandra Roy.
4. The text is given in Thibaut's article on Jyotisa Vedanga (*J.A.S.B.* 1877); *Yajusa Jyotisa* by Dvivedi and *Jyotisa Vedanga* by Lala Chotelal.
5. *Asiatic Researches,* vol. ix, p. 321.
6. Max Muller's *History of Ancient Sanskrit Literature.*
7. *Bhagavatisutra,* 90; *Uttaradhyayana sutra,* xxv, 7, 8, 38.
8. Rhys Davids, *Dialogues of the Buddha,* vol. II, pp. 20ff.
9. Cullavagga, viii, 6, 3, in Vinaya Texts. *Sacred Books of the East,* xx, pp. 292-4.
10. Gadaparva, 8, 14.
11. Salva parva, 38.

12. Thibaut, *Introduction to Pancasiddhantika,* p. xii.
13. *Ibid.,* p. xx.
14. *Ibid.,* p. xxv ff.
15. *Ibid.,* p. xxxviii.
16. P.C. Sen Gupta, Aryabhata—the father of Indian Epicyclic Astronomy, *Journal of the Department of Letters,* Calcutta University, Vol. XVIII.
17. Thibaut, *Introduction to Pancasiddhantika,* p. viii.
18. Lalla took 420 Saka as his karana year and as such he must have lived within 500 Saka at the latest. But Dikshit says that Lalla was born in 638 A.C. or 560 Saka; his argument are not convincing.
19. Vide a paper by Dr. B. B. Datta on the two Bhaskaras, *Indian Historical Quarterly,* Dec. 1930.

[Courtesy: Sukumar Ranjan Das]

9

Astronomical Instruments of the Hindus

"The principal instrument," says Brennand, "used by the Hindu astronomers was the Moon, which, from the rapidity of her motion, and the known places of the fixed stars on each side of her path, was an efficient means of determining the positions and motions of the planets by referring them directly to the nearest *Yoga-taras* of her path."

But to refer the moon to the nearest *Yoga-tara,* it was necessary to determine a point of the Ecliptic, namely the point of intersection of the Ecliptic with the Meridian. This point was called by the Hindus the *Kranti-pata,* the intersection of the Ecliptic with a circle of declination. The arc of the circle of declination from this point to the star, called the apparent latitude, and the arc of the Ecliptic from the same point to the beginning of the Asterism, *Asvini,* or to the first point of *Mesa,* called the star's apparent longitude, which were the necessary coordinates for the observations of the place of a star, required the help of some easy astronomical instruments; and also for observing during the day the sun's altitude and amplitude some instrument of the nature of the sun-dial had to be constructed.

This was apparent even at the time of the Vedas. Though the Vedic priest ascertained the motion of the sun by observing with his unaided eye the nearest visible star,[1] yet there is mention of the fact that Atri brought the sun back to view when enveloped in utter darkness by means of four Rks and the sons of Atri could alone release the sun from the engulfing darkness caused by a mighty power (राहु?)[2]. Tilak while discussing the meaning of the expression तुरीयेण ब्रह्मणा considers that Atri observed the solar eclipse by means of the quadrant instrument. Tilak's argument on this point is interesting and deserves special consideration. He says, "The fortieth hymn in the fifth Mandala of the Rg-veda is still more important in this connection. It shows that an eclipse of the sun was then first observed with any pretension to accuracy by Atri. It is thus that I understand the last verse in the hymn which, after describing the eclipse, says, "Atri alone knew him (sun), none else could".[3] This observation of the solar eclipse is noticed in the Sankhayana (24, 3) and also in the Tandya Brahmana (iv, 5, 2; 6, 13), in the former of which it is said to have occurred three days previous to the Visuvan (the autumnal equinox). The observation thus appears to have attracted considerable attention in those days. It seems to have been a total eclipse of the sun, and the stars became visible during the time, for I so interpret the expression *bhubanany adidhayuh* in verse 5. In verse 6 we are told that "Atri knew (the eclipsed sun) by *turiya brahma*", and Sayana interprets the last two words to mean "the fourth verse or mantra". But the verse wherein these words occur is to be understood the "fourth" if we count from the sixth, i.e. the tenth verse. The explanation may be good from the ritualist point of view, but it appears to me to be quite unsatisfactory otherwise. I would rather interpret *turiyena brahmana* to mean "by means of turiya." Turiya is mentioned in modern astronomical works as a name for an instrument called quadrant (Siddhantasiromani xi, 15), and though we may not suppose the same instrument to have existed in the old Vedic days, yet there seems to be no objection to hold that it may have meant some instruments of observation. The word *Brahman* is no doubt used to denote a mantra, but it may also mean knowledge or the means of acquiring such knowledge. In Rv., ii, 2, 7 Sayana has himself interpreted *Brahman* to mean some "act or action", and I see no reason why we should not understand the phrase *turiyena brahmana* in the above hymn to mean "by the action of turiya" or in other words, "by means of turiya," and thus give the whole hymn a simple and natural appearance."[4] There could be no other sensible explanation of the above two verses. Probably, Atri invented some instruments to observe eclipses whereby he could calculate the duration of full or partial solar eclipses and the last verse indicates that his descendants became conversant with observational astronomy by means of instruments.[5]

There is also mention of द्वादशाङ्गुल शङ्कु (Gnomon measuring 12 fingers) to measure the sun's shadow in the Atharva-veda.[6]

Then let us come to the text of the Jyotisa Vedanga. Therein are stated instruments like clepsydra and Gnomon to measure time and to find out the exact time.[7] In fact, the Hindus became so skilful in the use of clepsydra that they could find out the exact time at a mere look at the instrument. Probably the Gnomon was then used to determine the motion of the sun and other planets and to find out their exact position.

The Hindus at a very early date found out that without the help of suitable instruments the position of a planet could not be determined. The observation of the increase and decrease of the shadow of a tree must have struck them with the idea of a Gnomon. It could be easily constructed. The construction of a *cakra* or circle and a quadrant was very simple. It is, therefore, not impossible that the Hindus at the time of the Vedas used some instruments especially the quadrant to observe the eclipses. But though the Gnomon was sufficient to measure the time during the day, it was not possible to use it after dusk. The clepsydra was, therefore,

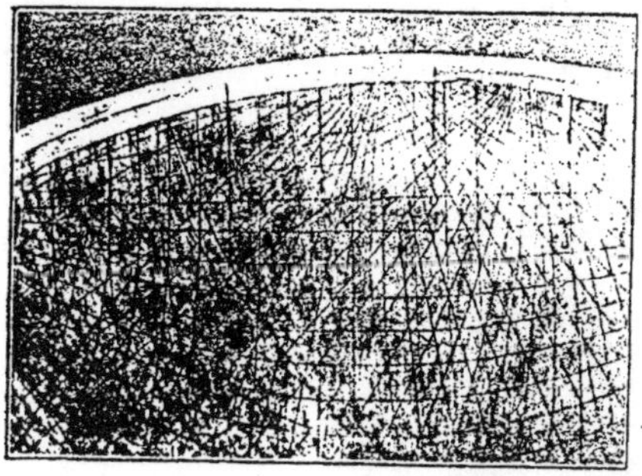

Kapila or Water-clock, Jaipur

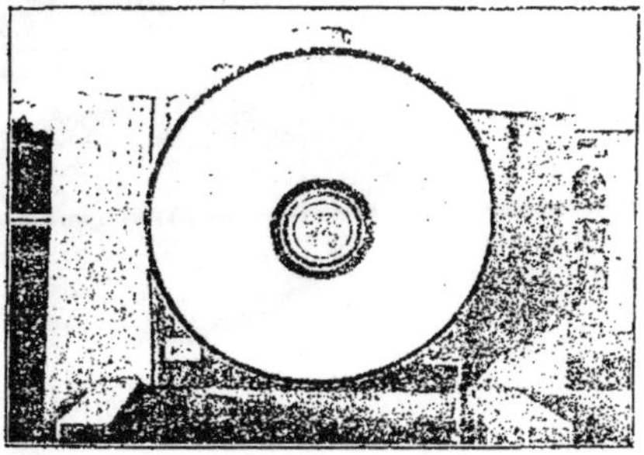

Nadi-valaya yantra, Jaipur

Astrolabe, Jaipur

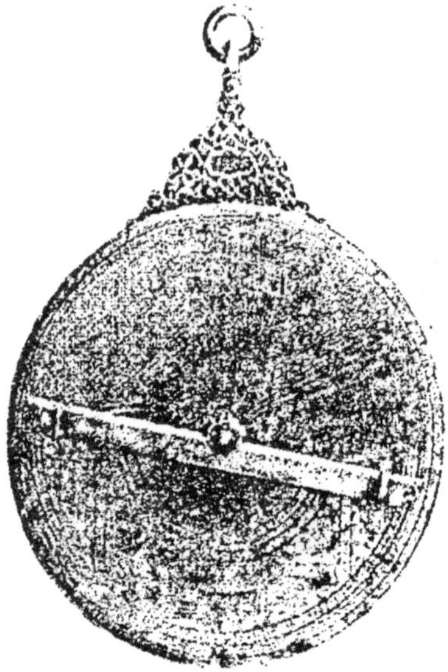

Back of the Astrolabe, Jaipur

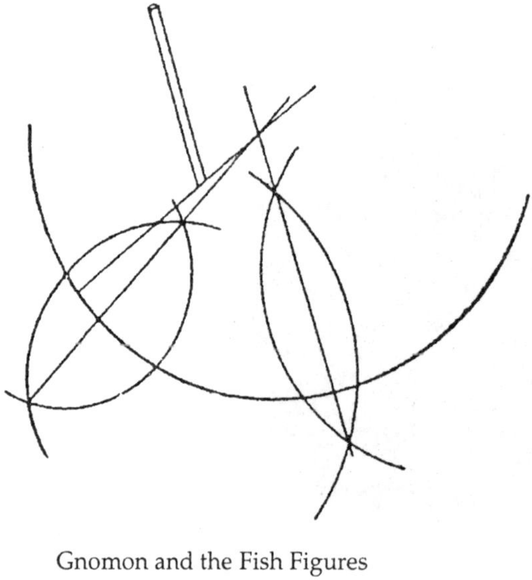

Gnomon and the Fish Figures

invented by the ancient Hindus. This instrument was in use till recently and it is even now found in some Hindu temples.

The period of the Jyotisa Samhitas of which the Brahma Samhita was the earliest came next. The main principles contained in the Samhitas are included in the Pancasiddhantika of Varahamihira. Hence we consider that the instruments mentioned in the Pancasiddhantika probably belong to the age of the Samhitas.

A systematic development of Indian astronomy began with Aryabhata in the fifth century A.C. (475 A.C.). But Aryabhata did not set apart a special chapter for describing the instruments. In the course of general discussion he makes use of the gola (sphere) and cakra (circle) instruments.[8] Let us now deal with the instruments found in the texts of the Siddhantas of later period:

1. Ghati or Kapala[9] (clepsydra or water-clock) is referred to in the Jyotisa Vedanga, where the amount of water that measures a nadika (24 minutes) is given.[10] The more ancient form of water-clock appears to have been simply a vessel with a small orifice at the bottom, through which the water flowed in a nadika,[11] but later on there came into use the form described in the Suryasiddhanta (chap. xiii, 23): "A copper vessel (in the shape of the lower half of a water jar) which has a small hole in its bottom and being placed upon clean water in a basin sinks exactly 60 times in a day and night." The description in Varahamihira's Pancasiddhantika[12] (505 A.D.) is similar, but adds, "or else a nadika may be measured by the time in which sixty slokas, each consisting of sixty long syllables, can be read out." The description of the water-clock given by Brahmagupta in his Brahmasphutasiddhanta[13] tallies with that given by the Suryasiddhanta. Lallacarya describes this instrument at a good length.[14] He further adds in his description of thc clepsydra:[15] "A copper vessel weighing 10 palas, 6 angulas in height and

twice as much in breadth at the mouth—this vessel of the capacity of 60 palas of water and hemispherical in form is called a ghati. The aforesaid copper vessel, bored with a needle made of 3 1/8 masas[16] (56 grains troy nearly) of gold and 4 augulas long, gets filled in one nadika." In practice, no doubt, the dimensions of the bowl and the orifice were determined by experiment. Rightly does Bhaskara say: "See how often it is filled and falls to the bottom of the pail of water on which it is placed. Divide 60 ghatis of day and night by the quotient and it will give the measure of the clepsydra."[17]

2. Sanku or gnomon is mentioned in the Atharva-veda as an instrument used to measure shadow. The sun-dial described in the early treatises is of the simplest kind, consisting of a vertical rod or gnomon, divided into twelve divisions. The Pancasiddhantika says: "Mark from the centre three times the end of the gnomon's shadow, and then describe two fish figures.[18] Thereupon describe a circle, taking for radius a string that is fastened to the point in which the two strings issuing from the heads of the fish figures intersect, and that is so long as to reach the three points marked. On the given day the shadow of the gnomon moves in that circle, and the base of the gnomon is the south-north line, and the interval, in the north direction, is the midday shadow."[19] The Suryasiddhanta remarks, "the gnomon is very useful by day when the sun is clear, and an excellent means of ascertaining time by taking its shadow."[20] The directions of the Suryasiddhanta are more elaborate. They are as follows:—"On the surface of a stone levelled with water or on the levelled floor of Chunar work, describe a circle with a radius of a certain number of digits. At the centre set up the gnomon of twelve digits of the measure fixed upon; and where the extremity of its shadow touches the circle in the former and after parts of the day, there, fixing two points upon the circle, and calling them the forenoon and afternoon points, draw midway between them, by means of a fish figure, a north and south line; midway

between the north and south directions draw, by a fish figure, an east and west line; and in like manner, also by the fish figures, between the four cardinal directions, draw the intermediate directions; draw a circumscribing square by means of the lines going out from the centre; by the digits of its base lines projected upon that is any shadow reckoned. The east and west line is called the prime vertical (samamandala); it is likewise denominated the east and west hour circle (unmandala) and the equinoctial circle (visuvanmandala); draw likewise an east and west line through the equinoctial shadow (visuvacchaya); the interval between any given shadow and the equinoctial shadow is denominated the measure of the amplitude (agra)". Brahmagupta describes the sun-dial thus:

Mule dvyangulavipulah sucyagro dvadasangulocchrayah/
Sankutalagraviddho' gravaidhalambad rjur jneyah//

Lalla gives a similar description and a more detailed one.[21] Of it Bhaskaracarya says, "take for a gnomon a cylindrical piece of ivory, and let it be turned on a lathe, taking care that the circumference is equal above and below; from its shadow may be ascertained the points of the compass, the place of the observer, including latitude etc., and times."[22]

3. The cakra or circle, marked on its circumference with 360°, is suspended by a string, the beginning of the divisions being at the lowest point. At the centre is a thin axis perpendicular to its plane. When the instrument is turned so that its plane is coincident with a vertical circle passing through the sun, the shadow of the axis is thrown on some division of the circumference and the arc between this point and the lowest point, the zero of the divisions measures the zenith distance or co-latitude of the sun. It is also used for finding the longitude of a planet; for if the instrument be inclined, and held or fixed so that any two of the stars Magha (Regulus), Pusya (Canceri), Revati, (Piscium), or Sata-taraka[23]

(Aquarii), appear to touch the circumference, the plane of the circle will coincide with the plane of the ecliptic, since these stars have no longitude. (Citra or Spica, whose longitude is inconsiderable and other stars near the Ecliptic would appear also to touch the circumference). The latitude of a planet which is generally small, has its orbit nearly in the same plane with that of the ecliptic. Looking, then, through a slit at the zero point of the circle, so that the planet appears opposite the axis, the position of the circle then remaining fixed, the eye is moved along the lower part of the circumference, so that any one of the above stars is seen opposite the axis, the arc between the two positions of the eye being the difference of longitude between the planet and the star; but the longitude of the star being known, that of the planet will also be known.[24] This instrument is also referred to in the Suryasiddhanta in a general way.[25] Brahmagupta in his Brahmasphutasiddhanta has devoted a couplet to the description of this instrument.[26]

Aryabhata[27] mentions this instrument and so does Lalla.[28] But Varahamihira has not mentioned it. However, this was a very simple instrument and it is believed that the Hindus used this instrument even at the time of the Vedas. Two such instruments can be seen in the Observatory at Jaipur (6ft. in diameter) and one at Benares (3ft 7 inches in diameter).

4. Capa—The half of a circle is called a capa. This instrument is mentioned only in the Siddhantasiromani.

5. Turiya or quadrant—The half of a semi-circle is called a turiya or quadrant. This is perhaps the oldest instrument mentioned by many old astronomers. I have already referred to its existence even at the time of the Vedas. But it is strange that Aryabhata, Lalla, Brahmagupta and the present text of the Suryasiddhanta do not take any notice of this instrument.

6. Pitha is a circle with an upward staff attached to it. It is mentioned only in Brahmasphutasiddhanta and Lalla's Sisya-dhivrddhida.[29]

7. Yasti[30] or staff is described at a considerable length in the Siddhantasiromani. Having drawn a circle (as the horizon) with a radius equal to the radius of a great circle, mark east and west points (and the line joining these points is called the pracyapara or east and west line) and mark off (from them) the amplitude at the east and west. Draw a circle from the same centre with a radius equal to a cosine of declination, i.e., with a radius of the diurnal circle, and mark this circle with 60 ghatis. Now take the Yasti, equal to the radius of the great circle and hold it with its point to the sun, so that no shadow be reflected from it; the other point should rest in the centre. Then measure the distance from the end of the amplitude to the point of the Yasti when thus held opposite to the sun. This distance applied as a chord within the interior circle will cut off, if it be before mid-day, an arc of the number of ghatikas from sun-rise, and if after mid-day an arc of the time to sun-set.[31] Brahmagupta makes a passing reference to this instrument. But Lalla describes it more fully.[32]

8. Dhanus or arc is only mentioned in the Brahmasphutasiddhanta and Siddhantasiromani.

9. Kartari is another instrument of the same type which has found place in Brahmasphutasiddhanta and Sisyadhivrddhida only.[33]

10. Bhaskara invented another instrument called Phalaka Yantra which is a development of the Cakra Yantra.[34] He says, "This is an excellent instrument, calculated to remove always the darkness of ignorance and is the delight of clever astronomers." This instrument is simply a board divided by horizontals into 90 equal parts which will be also digits. At the centre of the 30th graduation from the top of a pin or style is placed perpendicular to the board, and round it a circle drawn of radius =30 divisions, which is graduated in 60 ghatis and 360 degrees, each degree being subdivided into 10 palas; and attached to the pin is an index arm (pattika) with a

hole at one end and made of the length of 60 digits.[35] This instrument is suspended by a chain, and is used for observational purposes. It is in fact a kind of an astrolabe to be found in the Observatory at Jaipur.[36]

The gnomon (sanku), staff (yasti), arc (dhanuh), circle (cakra), half-circle (capa), quadrant (turiya), phalaka, pitha and kartari are instruments used for taking the shadow of various kinds.

11. A Svayamvaha or self-revolving instrument is thus described in the Siddhantasiromani: "Make a wheel of light wood in its circumference; put hollow spokes all having bores of the same diameter, and let them be placed at equal distances from each other; and let them also be all placed at angles somewhat verging from the perpendicular; then half fill these hollow spokes with mercury; the wheel thus filled will, when placed on an axis supported by two posts, revolve of itself."[37] The Suryasiddhanta also speaks of several self-revolving instruments used for measuring time. "A self-revolving sphere is to be made with its axis directed to the poles. The lower part of it is to be covered by wax-cloth, and it is to be made to rotate by the force of a current of water for the knowledge of the passage of time."[38] Also "a wheel with hollow spokes half filled with mercury, or water, or a mixture of oil and water will be made to revolve by itself."[39] Lalla's description of a self-revolving instrument is similar to that of the Siddhantasiromani. Probably, Bhaskara based his description of this instrument on that of Lalla[40] and developed it further.

In this connection Suryasiddhanta described many other unimportant instruments. It says, "By water instruments, the kapala (hemisphere) etc., by the peacock, man, monkey, and by stringed sand receptacles, one may determine time accurately. Quick-silver-holes, water, and cords, ropes and oil and water, mercury and sand are used in these. So also a dial (nara-yantra) is good in day-time, and when the sun is clear.[41]

12. Golayantra or Armillary sphere is described at a considerable length by all the ancient astronomers beginning from Aryabhata. Varahamihira, Lalla, Brahmagupta and Bhaskara paid much attention to this instrument. The Armillary Sphere was, however, of a nature too complicated to be used as an instrument for making accurate observations, and was rather for the purposes of explanation, and of giving instructions on the numerous circles and motions of the several spheres of which it was composed.

Generally it consisted of at least three separate spheres[42] on wood, on the same polar axis, or Dhruvayasti. "First, fix a celestial sphere named the Khagola, composed of circles for a given latitude, such as the horizon, the equinoctial, the meridian, the prime vertical, the six o'clock hour circle, the vertical circles through the north-east and north-west points of the horizon; the names of these circles are respectively called the Ksitija, the Nadi-vrtta (marked with 60 ghatikas), the Yamyottara-vrtta, the Sama-mandala, the Unmandala, the two Kona-vrttas with other circles, which always remain the same for the same place. Besides these fixed circles, a moveable altitude and azimuth circle is attached, by a pair of pins, to the zenith and nadir points of the Khagola, for showing the altitude or azimuth of any star. The horizon is divided in degrees, either from the meridian line or from the east and west points. Secondly, moveable within and round the axis of the Khagola was the starry sphere named the Bhagola, which comprised the Ecliptic, with the paths also of the moon and planets, named Ksepa-vrtta, the circles of declination, or Kranti, the diurnal circles called Ahoratra-vrttas; the azimuth circle through the nonagesimal point" (tribha lagna or tribhona) is called the Drkksepa-vrtta. The Bhagola is supported within the Khagola by means of two supporting circles called the Adhara-vrttas, corresponding with the meridian and horizon of the Khagola. Thirdly, on the axis of the Khagola produced, a third sphere is supported. It is called the Drggola, or Double Sphere, which is a system in which the circles of the Khagola are mixed with those of the Bhagola. The Khagola and Drggola remain fixed, while the Bhagola alone revolves."[43]

Suryasiddhanta gives instructions for the marking of an elaborate Armillary Sphere.[44] It says:—"Having fashioned an earth globe of wood, of the desired size, fix a staff, passing through the midst of it and protruding at either side for Meru; and likewise a couple of sustaining bands and the equinoctial circle; these are to be made with graduated divisions (angula) of degrees of the circle (bhagana). Further, by means of the several day radii, as adapted to the scale established for those circles, and, by means of the degrees of declination and latitude marked off upon the latter, at their own respective distances in declination, according to the declination of Aries etc., three bands are to be prepared and fastened; these answer also inversely for Cancer, etc. In the same manner, three for Libra, etc., answering also inversely for Capricorn, etc.; and, situated in the southern hemisphere are to be made and fastened to the two band supporters. Those, likewise, of the asterisms situated in the southern and northern hemispheres, of Abhijit, of the seven sages (saptarsis), of Agastya, of Brahma, etc., are to be fixed. Just in the midst of all, the equinoctial band is fixed. Above the points of intersection of that and the supporting bands are the two solstices (ayana) and the two equinoxes (visuvat). From the place of the equinox, with the exact number of degrees, as proportioned to the whole circle, fix by oblique chords the spaces (ksetra) of Aries and the rest; and so likewise, another band running obliquely from solstice to solstice, and called the circle of declination (kranti); upon that the sun constantly revolves giving light; the moon and the other planets also by their own nodes, which are situated in the ecliptic (apamandala), being drawn away from it, are beheld at the limit of their removal in latitude (viksepa) from the corresponding point of declination. The orient ecliptic point (lagna) is that of the orient horizon, the occident point (astam gacchat) is similarly determined. The meridian ecliptic point (madhyama) is as calculated by the equivalents in right ascension (lankodaya) for mid-heaven (hamadhya) above. The sine which is between the meridian and the horizon (ksitija) is styled the day-measure (antya), and the sine of the sun's ascensional distance (caradala) is to be recognised as the interval between the equator and the horizon. Having turned upward one's own place, the circle of the horizon is midway of the sphere. As covered with a casing (vastra) and as left uncovered, it is the sphere surrounded by Lokaloka. By the application of water is made ascertainment of the revolution of time."

Brahmagupta deals with the Armillary Sphere more in the Goladhyaya than in the Yantradhaya of his Brahmasphutasiddhanta. He says: pracyaparam samamandalam anyad yamyottaram ksitijam anyat parikaravat tanmadhye bhugolas tat-sthitadrastuh i.e. on it one circle called the samamandala or the prime vertical has its plane stretching north and south; another is termed the Yamyottara vrtta or meridian; yet another called the Ksitija or horizon encircles the other two like a girdle. At the common centre of these circles is the observer located on the earth (verse 48).

Also, "inasmuch as an observer on the earth finds a planet in the circumference of the upper half of the azimuth circle (drgmandala), it turns with the planet" (verse 55). Prthudaka Svami, the commentator of Brahmagupta, in his commentary, quotes at length a detailed account of the Golayantra or Armillary Sphere from earlier writers.[45] Lalla describes it in detail.[46] Bhaskara follows the description of Lalla very closely and says, "This instrument is to be made, placing the Bhagola starry sphere, which consists of the ecliptic, diurnal circles, the moon's path, and the circles of declination etc., within the

Khagola celestial sphere, which consists of the horizon, meridian, prime vertical, six o'clock line, and other circles which remain fixed in a given latitude. Bring the place of the sun on the ecliptic to the eastern horizon; and mark the point of the equinoctial (in the Bhagola) intersected by the horizon, viz., east point. Having made the horizon as level as water, turn the Bhagola westward till the sun throws its shadow on the centre of the earth. This distance from the mark made will represent the time from sun-rise. The Lagna or horoscope will then be found in that point of the ecliptic which is cut by the horizon."[47]

In this connection Bhaskara gives a brief description of another instrument which is called the Nadi-valaya,[48] a circle representing the Equinotial divided into ghatikas, and on it are the positions of the 12 signs, calculated to correspond with their oblique ascensions or risings at the place of observation. It is used in connection with the Khagola, whose axis casts a shadow on the circle, and is, in fact, an equatorial dial, the ghatika being 2/5 of an hour. Bhaskara describes it in the following verses:[49] "Take a wooden circle and divide its outer rim into 60 ghatikas; then place the twelve signs of the Ecliptic on both sides, but instead of making each sign of equal extent, they must be made each with such variable arcs as shall correspond with their periods of rising in the place of observation (the twelve periods are to be thus marked on either side, which are to be again each subdivided into two horas (or hours), three drekkanas, into navamsas or ninths of 3° i.e. 20 each, twelfths of 2° i.e. 10' and into trimsamsas or thirtieths These are called the sad-vargas or six classes). These signs, however, must be inscribed in the inverse order of the signs, that is 1st Aries, then Taurus to the west or right of Aries, and so on. Then place this circle on the polar axis of the Khagola at the centre of the earth (the polar axis should be elevated to the height of the pole). Now find the sun's longitude in signs, degrees etc. for the sun-rise of the given day (by calculation) and find the same degrees in the circle. Mark there the sun's place, turn the circle round the axis, so that the shadow of the axis will fall on the mark of the sun's place at sunrise and then fix the circle. Now as the sun arises, the shadow of the axis will advance from the mark made for the point of sun-rise to the nadir (adhah svastika) and will indicate the hour from sunrise and also the lagna (horoscope); the number of hours will be found on the shadow itself."

Colebrooke in his essay on the Indian and Arabian divisions of the Zodiac[50] gives a detailed account of the Armillary Sphere with the graduated circles, and remarks, "The instrument being thus placed, the observer is instructed to look at the star Revati, through a sight fitted to an orifice, at the centre of the sphere; and, having found the star, to adjust, by it, the end of the sign Pisces on the Ecliptic. The observer is then to look through the sight, at the Yoga-tara of Asvini, or of some other proposed object, and to bring the movable circle of the declination over it. The distance in degrees from the intersection of this circle and Ecliptic to the end of Mina or Pisces is its longitude, dhruvamsa in degrees; and the number of degrees on the movable circle of declination, from the same intersection to the place of the star, is its latitude, Viksepa, North or South."[51]

This is a detailed account of all the instruments used by the Hindus from very early times. The astronomical instruments of the observatories of Jai Singh have not been mentioned, though a passing reference has sometimes been made, inasmuch as they are very recent and belong to the eighteenth century A.D. Jai Singh studied Hindu, Muslim and European methods impartially and made use of both Hindu and Muslim instruments in his observatories. However, when everything is said, we must conclude by remarking that there is no living trace of most of the early astronomical instruments anywhere in India. It is desirable that Indian astronomers should try to revive the more important instruments by a careful study of the Hindu astronomical works.

References

1. *Tait. Br.*, i, 5.2, 1—Yat punyam naksatam tad basatkurvito-pavyusam, Yada vai surya udeti. Atha naksatram naiti, Yavati tatra suryo gacchet. Yatra jaghanyam pasyet. Tavati kurvita yatkari syat. Punyaha eva kurute. Tilak remarks, "The passage is very important as it describes the method of making celestial observations in old times."—Orion, p. 33.
2. *Rig-veda.*, v, 40, 6 and 9.
3. *Rig-veda.*, v, 40, 6.
4. Orion, pp. 159-160.
5. Vide Prof. Jogesh Chandra Roy's—*Our Astronomy and Astronomers*, p. 18.
6. अथर्ववेद विचार भाग, p. 18; vide also the discussion in Bharatiya Jyotihsastra by S. B. Dikshit, p. 367.
7. The clepsydra consists of a metal bowl floating in a vessel of water where the amount of water that measures a nadika (24 minutes) is given.
8. *Aryabhatiya*, Golapada, 22.
9. This is the instrument mentioned in the *Ain-i-Akbari* (ed. Jarrett), III, 16.
10. Lala Chhotte Lal's *Jyotisa Vedanga*, p. 12,
11. J.F. Fleet, The Indian Water-clock, *J.R.A.S.*, 1915, pp. 213. 230.
12. *Pancasiddhantika*, chap, xiv, 32,
13. *Brahmasphutasiddhanta*, Yantradhyaya, 41 and 42.
14. Lalla, *Sisyadhivrddhida*, Yantradhyaya, 10, 11.
15. *Ibid.*, 35.
16. Fleet quotes another rule giving the weight as a suvarna (16 masas) and length 4 angulas.
17. Siddhantasiromoni, Goladhyaya, chap, xi, verse 5.

18. *Timi* or fish figure is the space contained between two intersecting circles (see the figure).
19. Pancasiddhantika, p. 79; English translation, chap. XIV, 14-16, by Thibaut and Bapudev Sastri.
20. *Suryasiddhanta,* Chap. XIII, 24.
21. *Sisyadhivriddhida,* Yantradhyaya, 45-52.
22. Siddhantasiromani, Goladhyaya, Chap. XI, verse 9.
23. It is interesting to note that the mention of a hundred physicians in *Rv.,* 1, 24, 9 may be taken to represent the asterism Sata-bhisaj or Satataraka, presided over by Varuna according to the later lists of Naksatras in the Taittiriya Brahmana. Vide Orion, p. 159.
24. Siddhantasiromani, Goladhyaya, Chap. XI, 10-15.
25. Suryasiddhanta, Chap. XIII, 20.
26. Brahmasphutasiddhanta, Yantradhyaya, 18.
27. *Aryabhatiya,* Golapada, Arya 22.
28. *Sisyadhivriddhida,* Yantradhyaya.
29. *Sisyadhivriddhida,* Yantradhyaya, verse 25.
30. For the description of this instrument, vide S. B. Dikshit's *Bharatiya Jyotihsastra,* p. 348.
31. It is plain from this that the distance from the point of the staff to the end of the amplitude is the chord of the arc of the diurnal circle passing through the sun, intercepted between the horizon and the sun; for this reason, the arc subtended by the distance in question in this interior circle described with a radius of the diurnal circle which is equal to the cosine of the declination will denote the time after sun-rise or to sun-set (vide footnotes by Bapudev Sastri under verses 28-30, Siddhantasiromani, Goladhyaya, chap. xi).
32. *Sisyadhivriddhida,* Yantradhyaya, verses 48-50.
33. *Ibid.,* verse 24; also Brahmasphutasiddhanta, Yantradhyaya, verse 44.
34. Compare the description given by Delambre, *Astronomie du Moyen Age,* p. 521.
35. *Siddhantasiromani,* Goladhyaya, chap, xii, verses 18-22,

36. Vide the annexed figure.
37. *Siddhantasiromani,* Goladhyaya, chap. xi, 50-51.
38. *Suryasiddhanta,* chap, xiii, verse 16.
39. *Suryasiddhanta,* chap, xiii, verse 7.
40. Lalla, *Sisyadhivriddhida, Yantradhyaya,* verses 18, 19.
41. *Suryasiddhanta,* chap, xiii, verses 21-24.
42. See the annexed figure.
43. Brennand, *Hindu Astronomy,* p. 108.
44. *Suryasiddhanta,* chap, xiii, 3-16 (Burgess and Whitney's translation).
45. Vide notes in the chapter called Triprasnadhikara in Brahmagupta's *Khandakhadyaka* where a detailed description of Golayantra by Prthudaka Svamin has been quoted.
46. Lalla, *Sisyadhivrddhida,* Yantradhyaya, verses 3-6.
47. *Siddhantasiromani,* Goladhyaya, chap. vi.
48. This instrument is still found at the observatories of Jaipur, Ujjain and Benares. See the annexed figure.
49. *Siddhantasiromani,* B. D. Sastri's edition, chap. vi, 5-7.
50. *Asiatic Researches,* vol. 9.
51. Armillary sphere can be found at Jaipur and at Kotah.

[Courtesy: Sukumar Ranjan Das]

10

Greek Influence on Hindu Astronomy: Allegation or Reality?

Some writers like Mr. G. R. Kaye and Dr. D. E. Smith attribute the growth of Hindu Astronomy to Greek influence. This opinion was first expressed by Mr. G. R. Kaye and Dr. Smith, and others have followed suit. There must of course be considerable difference of opinion regarding the age of Hindu Astronomy in which personal equation of the writer must play a great part. Mr. Kaye assumes many things which perhaps take their colour from a preconception of superiority of the astronomical attainments of the Greeks to that of the Hindus.

European scholars chiefly divide Hindu Astronomy into two stages: that of the Vedas and that based on mathematical calculations as propounded in Brahmasiddhanta, Suryasiddhanata, Brihatsamhita, Siddhantasiromani and other astronomical works of more recent date. Baily, Max Muller and other eminent orientalists have expressed high admiration for the accurate astronomical knowledge attained by the Hindus even in the Vedic age; but a few oriental scholars of the school of Bentley observe nothing remarkable in the first stage of Hindu Astronomy, and Mr. Kaye, it seems to us, belongs to this school. Mr. Kaye remarks: "it is now acknowledged that the Hindu Astronomy of the second period (that of Aryabhata, Varahamihira, Brahmagupta, Bhaskara and others) came from the Greeks." He puts forward as his evidence that two at least of the western works, the Romaka and the Paulisa text books, were translated or adapted. Further, to prove that much of the astronomical knowledge came from the Greeks through the Persian medium, he adds that "there is the curious fact that the Hindus ascribed their knowledge of Astronomy to the Sun-worshipper Maya by whom they probably meant the Zoroastrian Ahura Mazda." The mention of the name of Maya puzzled Mr. Weber also who thought it was the Hindu translation of Ptolemois of the Greeks.

The basis of Hindu Astronomy lay in the religious aspirations of Hindu votaries in times when each heavenly body represented a Divinity. The study of astronomy originated in the doctrine that the Supreme Being had assigned duties to each of the heavenly bodies, by which they became rulers of the affairs of the world, and that a knowledge of the Divine Will would be acquired by watching and observing the order of their motions and the recurrence of times and seasons. With the Hindus the study of astronomy became a sacred duty, at least among the most educated classes, inasmuch as the celestial bodies were viewed as gods and the worship of them was enjoined by the Vedas. Thus the piety of the Hindus in primitive ages led them to watch with care all the phenomena of the heaven and to perfect their calendar of festivals, etc. To this end the first Hindu astronomers directed their attention. According to Baily, accurate astronomical observations had been made in India, probably before 3000 B.C.—a conclusion which is justified on independent evidence. Some writers of the history of the Vedic age have also held that the vedic sacrificial rites had a close connection with astronomical calculations. In any case, as they were regulated by the position of the moon with reference to the stars, they must be held to pre-suppose accurate astronomical observations, which indeed have come to be a religious necessity; it is therefore reasonable to argue, *a priori,* that an extensive astronomical knowledge was obtained in India even in the Vedic times. According to the Vedic hymns as translated by Pandit Satyavrata Samasrami, even in the Vedic times the Hindus had a knowledge of the motions of the planets (at least of five) and the causes of the solar and lunar eclipses. An astronomical interpretation of the Vedic hymns attempted by other scholars goes to show that a knowledge of solstitial and equinoctial points on the part of the Vedic writers could be reasonably accepted.

Mr. Weber suggests that Maya (as mentioned in *Suryasiddhanta*) is the Sanskrit translation of Ptolemois of the Greeks and thereby hints at the indebtedness of Hindu Astronomy to the Greeks. We shall just show how groundless this assumption is. The name of Maya appears in many places in the Puranas as a famous architect; and there is frequent mention of Mayavi-Maya (by this probably an eminent astrologer and astronomer is meant) in the Ramayana and the Mahabharata. Apart from the above, granting that Maya of the *Suryasiddhanata* is the Sanskrit counter-part of Ptolemois, we find that there is not a single passage in the whole of the *Suryasiddhanta* show that Maya gave any lessons on astronomy. He was only a learner and Surya is the preceptor. Surya is of course the god of the Hindus and was not imported from Greece. Therefore, even taking for granted Mr. Weber's interpretation, we draw just a contrary inference, viz., that Maya got his teaching from Surya. Again, the suggestion of Mr. Kaye that by Asura Maya is meant the Zoroastrian Ahura Mazda needs no refutation; for the mere perusal of any Persian treatise will convince any one that Persian Astronomy was not half so advanced as the astronomical knowledge evidenced in the Suryasiddhanta.

Of all the Sidhantas, none were held in such high esteem as the Brahma, Surya, Soma and Brhaspati. They were considered to be inspired. There were also two other

Siddhantas, Romaka and Paulisa. Mr. Kaye says that both of them are of Greek origin and have much influenced the ancient Hindu astronomers. The Romakasiddhanta is undoubtedly a translation of some Greek or Roman treatise as its name suggests. Its method of procedure also does not agree with that followed in any of the above Siddhantas and in this treatise the latitude of Alexandria has been adopted for calculating times and dates. Most probably it is an adaptation from Ptolemy's work, and its calculations have not been used, not even mentioned, in any of the known treatises of Hindu astronomy. The date of its composition has been placed by Dr. Kern in the sixteenth century, for in it there is an occasional mention of the name of Emperor Babar. Hence we may safely conclude that Romakasiddhanta has very little to do with the progress of astronomical knowledge of the Hindus. Of the Paulisasiddhanta we cannot say the same thing. Its method of procedure agrees in many respects with that of the current Siddhantas of the Hindus, the only difference being that the calculation of the solar and the lunar eclipses given in this book are not so accurate as propounded in the Suryasiddhanta or Siddhantasiromani of Bhaskara. It is believed by many European scholars that the Paulisasiddhanta is an adaptation of the work by Paulus Alexandrinus. But there was also a sage in India of the name of Pulisa. Is it not a faulty argument to base a conclusion on the similarity of names? "We have no right" says Dr. Kern in his preface to Brhatsamhita, "whatever, to infer that Paulus Alexandrinus and Pulisa are one and the same, for identity of name is too slender a ground specially when the name happens to be a common one." Further, the Paulisasiddhanta (as described by Prof. Jogesh Chandra Ray in his work entitled "Our Astronomy and Astronomers") is based on mathematical calculations, whereas the work of Paulus is mainly an astrological one. We have, therefore, every reason to believe that Paulisasiddhanta had its origin in India and is not imported from outside.

The second period of Hindu Astronomy was a real advance in mathematical astronomy and the reasonings set forth there exhibit a keenness of observation that would do credit to latter-day European philosophy. The treatises of that period that are held in high esteem by the Hindus are Brahmasiddhanta, Suryasiddhanta, Brhatsamhita and Siddhantasiromani. The dates of composition of the above works have not been fixed to our entire satisfaction; there exists a vast difference of opinion as to the actual dates among the oriental scholars. As regards dates we are inclined to follow Mr. Baily, Mr. Brennand and Pandit Sudhakara Dvivedi where any difference of opinion arises.

The birth of Aryabhata marked a new era in the progress of mathematical astronomy. He wrote a number of works on astronomy which are now known only through quotations from his writings given by Brahmagupta and other subsequent astronomers. It is, in general, through these citations that Aryabhata was known to be an eminent astronomer anterior to Brahmagupta in the beginning of the sixth century A. C. From the quotations of Brahmagupta we learn that Aryabhata believed in the diurnal motion of the earth round its axis. "The starry sphere" he affirms "is stationary, and the earth making a revolution produces the daily rising and setting of stars and planets." The idea of heliacal risings and settings of stars and planets was first fully developed in Europe by Copernicus in the fifteenth century, and before him, it was not recognised though some rudimentary hints were thrown by Pythagoras in the fifth century B.C. Aryabhata promulgated the above theory in the sixth century A.C. at the latest. Hence we see that Mr. Kaye's statement that the Hindus borrowed the idea of heliacal risings and settings of stars and planets from the Greeks does not stand scrutiny. We are rather inclined to believe that this theory had its origin in India and was introduced in Europe through the Greek medium, by which it was given its present practical garb. The confusion as regards the correct date of Aryabhata arises from the fact that most of the oriental scholars have failed to notice that there were two men of the name of Aryabhata—one, the great astronomer who lived in the fifth century A.C. and the other, a mere commentator and compiler, who flourished in the thirteenth or fourteenth century A.C. The latter wrote a commentary on Suryasiddhanta. Dr. Bhau Daji says that the work mentioned by Alberuni as written by Aryabhata is this commentary. If this be true we can infer that Suryasiddhanta in its present form was current before the ninth century; but we have evidence enough to show that some portions of the Suryasiddhanta were written even before the age of the Sulva-sutras.

The name of Brahmagupta is held in high esteem by all Hindu writers. He flourished in the sixth century A.C. We know that there is not a single argument in the whole range of Greek Astronomy to prove why the earth, though spherical, appears flat to observers on earth. "The earth," says Brahmagupta, "stands firm by its own power without any other support in space. If there be a material support to the earth and another upholder of that and again another of this and so on without limit, and if finally self-support must be assumed, why not assume it in the first instance? Why not recognise it in this multiform earth? The earth possessing an attractive force draws towards itself any heavy substance situated in the surrounding atmosphere, and that substance appears as if it fell (like a stone). But whither can the earth fall in ethereal space which is equal and alike on every side? ...... If the earth were falling, an arrow shot into the air would not return to it when the projectile force was expended, since both would descend; nor can it be said that the earth moves slower and is overtaken by the arrow, for heaviest bodies fall quickest and the earth is the heaviest." Such reasonings can do credit even to modern astronomers.

Again why does not the earth appear spherical to men on earth? Aryabhata in reply says:

"As the earth is a large body, and a man is exceedingly small, the whole visible portion of the earth consequently appears to a man on its surface, to be perfectly plane." We think this is a point gained by Hindu Astronomy over the Greek.

Next came Barahamihira who also lived in the sixth century A.C. He was not an original writer but a compiler. One of the passages in his work, Brhatsamhita, has led Mr. Kaye to think that Baraha admits that the Hindus have borrowed much from the Greeks. We shall presently show that he has misunderstood and misinterpreted the couplet referred to. He translates the couplet thus—"The Greeks, indeed, are foreigners, but with them this science is in a flourishing state. Hence they are honoured as though they are Rsis." The couplet as written by Baraha is this:

Mleccha hi yavanas tesu samyak sastram idam sthitam /

Rsivat te' pi pujyante kim punar daivavid dvijah //

The translation should be as follows: "The Greeks are foreigners, with them also this science is in a flourishing state. Even they are honoured as Rsis; much more will be honoured a Brahmin versed in astrology." Scholars without sufficient knowledge of Sanskrit are likely to advance this couplet as a proof that the Hindus learned astronomy from the Yavanas (Greeks), but one should not fail to take notice of the fact that the above couplet occurs in the astrological section of Brhatsamhita and also in connection with the word "Daivavid," meaning astrologer and not astronomer. From the Brhatsamhita edited and translated by Pandit Sudhakar Dvivedi we find that the names of Yavana (Greek) teachers occur sixteen times and everytime in support of the views of Baraha as regards "lagnasuddhi" and "varasuddhi" and at no time in support of astronomical facts.

The next point that engages our attention is the explanation for the planetary motions by the method of epicycles. Mr. Kaye argues that this was of Greek origin and the Hindus borrowed it from the Greeks. The first notion of the planetary motion was clearly stated in the first chapter of Suryasiddhanta, and this, we are led to believe from the citations of ancient astronomers, existed in the oldest edition of Suryasiddhanta, the date of which can be placed somewhere near the second or third century A.C.

"The planets in their orbits go rapidly and continually with the stars towards the west, and hang down at an equal distance as if overpowered (over-matched in speed) by the stars. Therefore, the motions of the planets appear towards the east, and their daily motions, determined by their revolutions, are unequal to each other in consequence of the circumstances of their orbits; and by this unequal motion they pass the sign. The planet which moves rapidly requires a short time to pass the signs, and the planet that moves slowly passes the signs in a long time" (*Suryasiddhanta,* chap. I, slokas 25-27).

This notion that the motion of all the planets was caused by a velocity in their orbits, which was the same for all alike, was prevalent not only in India but also in Europe even to the times of Kepler and Newton. This is evident from the manner in which Kepler combated this doctrine and from the important use he made of it. Soon after the death of Tycho, Kepler "made many discoveries from Tycho's observations. He found that astronomers had erred from the beginning of the science in ascribing circular orbits to the planets. He easily saw that the higher planets not only moved in greater circles but also more slowly than nearer ones, so that on a double account their periodic times were greater."

This planetary motion the Hindus and the Greeks explained by means of epicycles. Aryabhata ascribed to the epicycles by which motion of a planet was represented a form varying from the circle nearly to an ellipse. It was well-known to the Hindus that a supposed uniform motion in a circle about the earth did not really represent true motion in its orbits, although the hypothesis served sufficiently to determine the mean motions and the mean place of a planet when deduced from observations carried on for lengthened periods. They knew that every planet in its course was subject to great irregularities, the motion undergoing continual changes. At one time it would be direct towards the east until the planet reached a stationary point where it would seem to be at rest; then a retrograde motion would begin and continue for a time till another stationary point was reached, and the eastward motion was repeated. It was to account for these irregularities that the epicycle was invented. By the Greeks this contrivance was ascribed to Apollonius. He conceived that a planet in its course, described with uniform motion and the circumference of a circle, is called the epicycle, whose centre moved uniformly in the circumference of another circle, called the deferent, the centre of which was the centre of the earth. It was also supposed that, whilst the centre of the epicycle was moving eastward in the direction of the signs, the planet itself was moving in a direction opposite to that of the sign. By this hypothesis it was easy to show the various changes in the motion of the planets. This theory was generally adopted by western nations with the addition of other epicycles, introduced by Ptolemy, necessary for expressing the apparent motions with accuracy. But the Hindus had two different methods for calculating the true place of a planet from its "mean place" as determined by the rules of *Siddhantasiromani* (chap. II). One of these methods resembled that of Apollonious, but there was also considerable difference, viz., that whilst the planet moved uniformly in its epicycle, whose centre moved in the deferent concentric with the earth, the epicycle itself was conceived to be variable, the circumference being the greatest when the planet was in an apsis (at Apogee or Perigee the true and mean places being then coincident), and the least when the planet was at a distance of 90° from the points. The other method supposes that, while the mean place of a planet is a point moving uniformly eastward round the circumference of an equal but eccentric circle, whose centre is situated in the line joining the Apogee with the centre of the earth, the distance from it is the eccentricity. These two methods of calculation, whether by assuming the motions as being in an eccentric or in an epicycle, give exactly the same results. The theories of epicyclical and eccentric motion of planets are beautifully put in the second chapter of Suryasiddhanta and the fifth chapter of Siddhantasiromani. A lengthy comparison of the Hindu and Greek methods together with the remarks of Bhaskara by means of figures and mathematical calculations will require an independent paper on the subject. But we hope that the main point of the Greek and Hindu epicyclical theories that have already been mentioned will convince one that unlike the epicycles of Ptolemy and other Greek astronomers, the Indian epicycles had a variable circumference, that of the first epicycle being the largest at Apogee and Perigee, varying from those points through the deferent to its places at the quardrants, where its circumferences were the least. Now we believe we may be permitted to draw the conclusion that the two different methods of the Hindus and the Greeks grew up side by side without the one being influenced by the other; for astronomical arguments point to a possibility, if not a certainty, of this kind of coincidence.

Next in importance comes the question of discovery of the twelve signs of the Zodiac. Mr. Kaye has taken it for granted, *a priori,* that the division originated in Greece and thence was borrowed by the Hindu astronomers. We shall show that astronomical arguments lead to the very opposite conclusion. Here we shall follow the admirable treatment of the question by Prof. Dr. D. N. Mallik in the course of one of his lectures: "The Hindu astronomers had two systems of reckoning: the lunar mansions or the *Tithis* and the signs of the Zodiac or the *Rasis,* the first being obviously the earlier of the two. For, while the moon's motion among the stars is a matter of direct observation, the solar motion in its relation to the stars could only be observed by an indirect method, on account of the fact that his light shuts out of view all stars in his neighbourhood. On the other hand, the moon's motion is much more irregular than that of the sun. The observation of the sun's motion, therefore, came gradually to be recognised as a matter of practical as well as of scientific importance and the method of signs or Rasis ultimately superseded that of the Tithis. As to the lunar system of the Hindus, its high antiquity is testified to by the fact that the primitive series opened with Krittika (the Pleiades) as the sign of the vernal equinox. But this arrangement would be correct only about 2300 B.C. and nowhere else would be found a well-authenticated Zodiacal sequence of so early a date. If this be granted, it seems to be very probable that the method of signs was evolved in India, for the method of tithis which is admitted to be peculiar in India may be regarded as the parent of the method of signs, and we are thus almost able to trace a gradual evolution along lines well-recognised by the science of the system of signs." The Hindu astronomers divided the Ecliptic and the Zodiac into 28 parts (and then into 27) forming so many groups of stars in the path of the moon, each division corresponding with the space of the moon's daily motion through them. The groups were consequently called lunar asterisms. The ancient Hindu astronomers chose a set of 27 principal stars, one for each of the 27 lunar constellations, in general the brightest star of the asterism, and called it Yoga-tara, whilst the asterism-cluster was named the Naksatra. The Yoga-tara was connected with the beginning or first point on the Ecliptic of the division, representing the space of the asterism by the small arc of apparent difference of longitude between them, this arc being called the Bhoga of the asterism. Thus the 27 divisions of the ecliptic became as fixed in position as the stars themselves, like a great fixed dial, with the numbers ranging not along the Equator but along the Ecliptic itself.

From the above arguments, and from investigations of Prof. Weber and Colebrooke we deduce that the Hindus founded their lunar mansions which the Arabs (by the name of Manzil) and the Chinese (by the name of Sieu) borrowed. We also know that the Greeks were not possessed of the system of lunar mansions. Prof. Weber, however, propounded the view that the system originated in Babylon. Such a view can no longer be maintained, since we have now to admit that the Babylonian system is based on the sun's motions. Now if the view advocated above that the lunar system must be of older date than that which is based on the sun's motions be correct, we must admit that the Babylonian system must needs have been derived by adoption. Hence we are justified in inferring that if the Hindus can claim to be the originators of the system of lunar mansions, they have an equal right to claim to be the propounders of the system dependent on the division of the Zodiac.

The next point that engages our attention is the question of parallax. Mr. Kaye says that the idea of parallax in Hindu Astronomy is also borrowed from the Greeks. We are not prepared to accept his theory which is not supported by reasons. The Hindus, we know, were at a very early date well-acquainted with all facts relating to eclipses. They had rules of calculating the various phases of both the lunar and solar eclipses, i.e., the times of their beginning, middle and end which are set forth in the various astronomical works. The correction of parallax in latitude and longitude is of great importance in calculating eclipses accurately and hence we are led to believe that the Hindus had a thorough knowledge of this phenomenon even in the Vedic ages when eclipses were calculated for religious purposes. Bhaskara in his *Siddhantasiromani* quotes several couplets from ancient astronomers explaining the importance of the correction of parallax in calculating eclipses. Barahamihira in his *Panchasiddhantika* also elaborately deals with it under a section of Suryasiddhanta.

These are some of the points that present themselves to one engaged in studying Hindu Astronomy. The remarks of many an oriental scholar that the Hindus have not received the credit due to their astronomical science and the publications of Mr. Kaye and Dr. D.E. Smith have led me to adduce some proofs of the advanced stage of astronomical knowledge reached by the Hindus without any outside aid. We hope, however, that every one will admit that there are many points in Hindu astronomy which for their scientific importance must rank first in the history of the science.

[Courtesy: Sukumar Ranjan Das]

Index

❑❑❑